PESTICIDE REMOVAL BY COMBINED OZONATION AND GRANULAR ACTIVATED CARBON FILTRATION

Pesticide Removal by Combined Ozonation and Granular Activated Carbon Filtration

DISSERTATION
Submitted in fulfilment of the requirements of
the Academic Board of Wageningen University and the
Academic Board of the International Institute for Infrastructural,
Hydraulic and Environmental Engineering for the Degree of DOCTOR
to be defended in public
on Monday, 22 November 1999 at 15:30 h
in the Auditorium of IHE-Delft

by
ERVIN ORLANDINI
born on 3 February 1962 in Split, Croatia

A.A. BALKEMA / ROTTERDAM / BROOKFIELD / 1999

This dissertation has been approved by the promoter:

Prof. dr ir J.C. Schippers   Professor of Water Supply Technologies at Wageningen University and the International Institute for Infrastructural, Hydraulic and Environmental Engineering

Published by
A.A. Balkema, P.O. Box 1675, 3000 BR Rotterdam, Netherlands
Fax: +31.10.4135947; E-mail: balkema@balkema.nl; Internet site: http://www.balkema.nl

A.A. Balkema Publishers, Old Post Road, Brookfield, VT 05036-9704, USA
Fax: 802.276.3837; E-mail: info@ashgate.com

ISBN 90 5410 414 7

Printed in the Netherlands

Contents

Acknowledgments

First, I thank my promoter Prof.dr.ir. Jan Schippers for providing valuable advice throughout this research. His broad view, not just limited to issues related to water supply technologies, and meticulous –very Dutch– analysis of every statement made in this thesis made an lasting impression.

Dr. Joop Kruithof always found time to share his vast knowledge of various aspects of this research, and ensured a friendly safeguard against possible mistakes. Prof.dr. Vernon Snoeyink and Prof.dr.ir. Guy Alaerts, who together with dr. Kruithof formed the Reading Committee, gave the most valuable comments on the draft version of this thesis.

This research was done within the project conducted jointly by IHE, NORIT NV, Kiwa NV and Amsterdam Water Supply, and was also supported by the Dutch Ministry of Economic Affairs (Senter grant no. MIL93042). The members of the Project Committee – ir. Jos Boere, dr. Ton Graveland, dr.ir. Jan Peter van der Hoek, drs. Rob de Jonge, dr. Maria Kennedy, dr. Joop Kruithof, Prof.dr.ir. Jan Schippers and dr.ir. Maarten Siebel – contributed to the development of the initial hypotheses and discussions of the results obtained.

Contributions of many that were involved in the experiments and analyses conducted are acknowledged in the corresponding thesis' chapters. Nevertheless, I would like to mention Ada Vooijs here, and to thank her for her accuracy and persistence while analyzing many atrazine samples collected. I also thank Afshan Shafi for proofreading my English.

Last but not the least, I thank Olja and Lara for the joy they give, and Rosana for her confidence and love.

Abstract

Orlandini E. (1999). *Pesticide removal by combined ozonation and granular activated carbon filtration.* Ph.D. thesis, International Institute for Infrastructural, Hydraulic and Environmental Engineering (IHE) and Wageningen University, 171 pages.

Since the seventies, new water treatment processes have been introduced in the production of drinking water from surface water. Their major aim was to adequately cope with the disinfection of this water, and/or with the removal of pesticides and other organic micropollutants from it. Amsterdam Water Supply (AWS) recently developed two new integral concepts for the treatment of Rhine River water. They involve conventional pretreatment of directly taken river-water by coagulation, sedimentation and rapid sand filtration, followed either by slow sand filtration and reverse osmosis, or by ozonation, Granular Activated Carbon (GAC) filtration, slow sand filtration and reverse osmosis.

In contrast to water treatment schemes typically applied in The Netherlands, these two concepts do not use a reservoir or artificial recharge to provide for the periods when the Rhine River is heavily polluted, or to improve the quality of water. Thus, they have to offer so robust barrier against all possible pollutants in Rhine River water that under both normal and accidental conditions safe drinking water can be produced. AWS also aims to reduce the salinity of the raw water, in particular the chloride concentration, and to continue production of soft and biologically stable water. AWS tested and evaluated the performances of these two concepts for the future capacity extension at its Leiduin plant. The mechanisms that play a role were studied in the context of a research project conducted jointly by IHE, NORIT NV, Kiwa NV and AWS. The research presented in this thesis was done within the framework of this project. Its focus is on Biological Activated Carbon (BAC) filtration, which is a combination of ozonation and GAC filtration.

The general goal of this research is to identify and understand mechanisms that underlie the expected beneficial effect of ozonation on the removal of organic micropollutants by GAC filtration. This understanding allows one to judge whether this combination provides a sound barrier against these compounds and, in addition, it allows optimization of the underlying mechanisms. Detailed investigations for all organic micropollutants present in Rhine River water were not possible, thus, one model compound was chosen. Pesticide atrazine was chosen as a model compound because, at the start of this research in 1992, its removal was particularly relevant for AWS; it was detected in pretreated Rhine River water at concentrations higher than the European Union standard of 0.1 μg/l. Moreover, the analytical method needed to measure atrazine concentrations below 0.1 μg/l was available.

Pilot plant experiments with Rhine River water (pretreated by coagulation, sedimentation and rapid sand filtration) confirmed the expectation that ozonation significantly improves the removal of atrazine by GAC filtration. This improvement is not only due to the well-known effect of ozone-induced oxidation of atrazine, but also due to the effect of ozone-induced oxidation of a part of Background Organic Matter (BOM) present in water. BOM refers to the organic matter in the influent of GAC filters other than the target compounds that need to be removed. BOM is mostly of natural origin, *e.g.* compounds such as humic substances, but it also includes –especially in Rhine River water– many compounds of anthropogenic origin.

The scope of this research was limited to the investigation of the removal of atrazine in GAC filters. The removal of the by-products of atrazine oxidation formed by ozonation (*e.g.* desethylatrazine and desisopropylatrazine) was thus not investigated. Considering that these compounds are expected to be more biodegradable and (as shown in this research) less adsorbable than atrazine, it is difficult to predict whether they are removed by GAC filtration better or worse than atrazine.

Subsequent (pilot-, bench- and lab-scale) experiments aimed to verify which of the anticipated processes and relationships underlie the improved atrazine removal observed in filters receiving ozonated influent. Namely, an important part of BOM compounds is partially oxidized due to ozonation. This partial oxidation increases the biodegradability, and reduces the adsorbability and molecular mass of BOM compounds. Consequently, enhanced biodegradation of BOM and its reduced adsorption are expected in filters receiving ozonated influent. Both biodegradation and adsorption of pesticides are expected to be improved as a result.

These experiments led to the following conclusions. Improved removal of BOM observed in filters that received ozonated influent can be attributed to the enhanced biodegradation of BOM in these filters. This can be concluded because ozonated, rather than non-ozonated BOM, was also better removed in filters filled with non-activated carbon, in which the removal of BOM is via biodegradation only. It could not be demonstrated that biodegradation of atrazine accounts for its improved removal in GAC filters that received ozonated rather than non-ozonated influent. Namely, no indication of atrazine biodegradation in these GAC filters was found in either of the experiments conducted: (i) no metabolites of atrazine were detected in the effluent of atrazine-spiked GAC filters, (ii) atrazine was not removed in filters filled with non-activated carbon, (iii) atrazine was not removed in the liquid media inoculated with the bacteria taken from atrazine-spiked GAC filters, and (iv) after three years of pilot plant operation, more atrazine was desorbed from GAC taken from the filter that received ozonated rather than non-ozonated influent. The enhanced biodegradation of BOM in filters receiving ozonated influent improves adsorption of atrazine in GAC filters. This can be concluded because atrazine was better adsorbed onto GAC preloaded with ozonated water that passed through filters filled with non-activated carbon (in which the removal of BOM is via biodegradation) than onto GAC preloaded directly with ozonated water.

The results also revealed that the improved adsorption of atrazine in filters receiving ozonated influent is the effect of both the higher adsorption capacity of GAC for atrazine, and the faster external and internal mass transfer rates of atrazine in these filters compared with filters receiving non-ozonated influent. Higher adsorption capacity and faster mass transfer were explained as due to reduced competitive adsorption and reduced preloading of ozonated BOM. Competitive adsorption of BOM occurs when BOM adsorbs simultaneously with atrazine, and competes with it for the adsorption sites available on GAC. BOM preloading is adsorption of BOM onto GAC before the adsorption of atrazine. Reduced competitive adsorption and reduced preloading of ozonated BOM are the consequence of increased biodegradability of a part of BOM compounds that are partially oxidized by ozonation. Namely, this increases the amount of BOM that is biodegraded rather than adsorbed in GAC filters. Besides increased biodegradability, decreased adsorbability of oxidized BOM also contributes to the improved adsorption of atrazine.

Finally, the two commonly applied models, *i.e.* the simple Adams-Bohart model and the more complex Plug Flow Homogenous Surface Diffusion model, were applied for the prediction of atrazine breakthrough in GAC filters with and without ozone-induced bioactivity. Neither model resulted in an accurate prediction. This can be expected considering that, due to the complexity of the processes that simultaneously take place during GAC filtration, the prediction of its performance involves many inevitable assumptions and simplifications.

Key words: activated carbon, adsorption, atrazine, Background Organic Matter, biodegradation, biological activated carbon filtration, bromate, competitive adsorption, disinfection, modeling, ozonation, pesticides, preloading, process analysis.

Chapter 1

General Introduction

ABSTRACT—Disinfection and removal of organic micropollutants are the two aspects of the production of drinking water from surface water that are currently of particular importance. They are important issues because conventional water treatment processes are not always able to cope with them. This was first realized when chlorine, a commonly used disinfectant, was found to result in the formation of trihalomethanes, and when various micropollutants were detected in drinking water at concentrations far exceeding acceptable levels. Thus, new, advanced water treatment processes are needed.
Amsterdam Water Supply (AWS) introduced two new integral concepts for the treatment of surface water. They include conventional pretreatment of Rhine River water by coagulation, sedimentation and rapid sand filtration, followed either by slow sand filtration and reverse osmosis or by ozonation, Granular Activated Carbon (GAC) filtration, slow sand filtration and reverse osmosis. The performances of these two concepts were tested and evaluated by AWS. The mechanisms playing a role were studied in the context of the research project conducted jointly by IHE, NORIT NV, Kiwa NV and AWS. In particular, attention was paid to the following aspects: removal of pesticides, metabolites and other organic micropollutants by combined ozonation and GAC filtration, and by reverse osmosis; disinfection by ozonation and by reverse osmosis; and control of the fouling and scaling of reverse osmosis membranes.
The research presented in this thesis was conducted within the framework of this project. Its focus is on the removal of pesticides by Biological Activated Carbon filtration, which is a combination of ozonation and GAC filtration.

1.1 TREATMENT OF SURFACE WATER

1.1.1 History

The earliest recorded knowledge of water treatment can be found in *Sus'ruta Samhita,* a collection of Sanskrit medical lore thought to date from 2000 B.C. It declares that impure water should be purified by boiling, or by exposing to sunlight, or by filtering through sand and coarse gravel and allowing to cool. Other ancient civilizations also left behind some evidence of their water treatment practices. Egyptians siphoning off water (or wine!) clarified by sedimentation were pictured on the wall of a tomb built at Thebes in 1450 B.C. A somewhat less sophisticated method was the drinking cup devised by the Spartan ruler Lycurgus (ninth century B.C.), which hid badly colored water from the sight of the drinker and allowed mud to stick to its side. Cyrus the Great, King of Persia in the sixth century B.C., was known to take boiled water in silver flagons along with him when going to war. In this way, water sterilized by boiling was kept sterile through the germicidal action of silver. The simile in Plato's *Symposium* (fourth century B.C.) suggests that Greeks commonly used wick siphons to clarify water. In this simile, Socrates says that it would be a good thing if wisdom could flow from a person full of it to a person less wise, just as water flows through a thread of wool from a fuller to an emptier vessel (Baker, 1949).

The first reference to public water supply was made by the Roman engineer Sextus Julius Frontinus, who in 97 A.D. became water commissioner of Rome. One year after his appointment as water commissioner, he wrote the first known detailed description of water works systems: *De Aquis Urbis Romae Libri II.* Among others things, these two books describe the *piscanae* or pebble catchers built into most aqueducts to serve the double purpose of storing and of clarifying the water (Baker, 1949).

Thus, throughout history, the greatest concern regarding the use of surface water as a source of drinking water was to ensure its clarity and microbiological integrity. Water was clarified by sedimentation and/or filtration, neither of which was quite satisfactory. While easy to operate, settling basins result in an effluent of rather low clarity. Better clarity of water is obtained by filtering it over a coarse material like gravel; however, over time, clogging occurs and filters need to be cleaned. Until one and a half centuries ago, cleaning of the filter was laborious and time-consuming: filter material had to be taken out from the filter, washed, and then put back into the filter. A new way of ensuring the clarity of surface water, superior to either sedimentation or filtration, came into use in England in the first decade of the 19th century. The new process, termed "slow sand filtration", combined the use of fine filtering material and low filtration rates. The fine material limited clogging to the initial few centimeters of the filter bed, which can be easily scraped away, while the low filtration rates extended the interval between two scrapings to a few months. In 1829, James Simpson built the first slow sand filters used for a public water supply, the Chelsea Water Company in London.

The biological/physical process of slow sand filtration improves not only the clarity but also the microbiological quality of water. This was first indicated in 1849, when much higher incidences of cholera were noted in the districts of London where water was clarified by sedimentation than in the districts where slow sand filtration was used. The final and clear evidence was provided in 1892, when Hamburg experienced a massive outbreak of cholera, while there were only a few cholera cases in the neighboring city of Altona. Both cities used the Elbe River water as a source of drinking water. In Hamburg, however, it was treated by sedimentation, while in Altona it was treated by slow sand filtration (Huisman, 1990).

Due to the simplicity of its design and operation, and the quality of the water produced, slow sand filtration is still used in both industrialized and developing countries. Over time, it has been extended by coagulation, sedimentation and rapid sand filtration as a pretreatment for the removal of suspended matter and algae, and with post-chlorination for additional disinfection.

After World War II, many water supply companies set up a purely physicochemical system for the treatment of surface water. The new concept required less space than the one based on slow sand filtration, and consisted of pre-chlorination, coagulation, sedimentation, rapid sand filtration and post-chlorination. These basic concepts are frequently extended with the storage of raw water in reservoirs or in the underground. This is done to improve the quality of water by self-purification and equalization, and to allow no intake when the quality or the quantity of raw water is exceptionally low. The abstraction of bank filtered water, and the addition of powdered activated carbon to remove taste and odor, are also frequently applied.

By the end of the sixties, these conventional water treatment processes[1] seemed to suffice for the treatment of surface water. Opinion changed drastically with the finding that chlorine reacts with humic substances present in surface water to form trihalomethanes (Rook, 1974). In addition, at around the same time, there was growing awareness that many surface waters were polluted with potentially hazardous organic chemicals of industrial origin that could not be effectively removed. This started the ongoing discussion about the ability of water treatment processes to provide adequate disinfection of surface water, and a sufficient barrier against organic micropollutants present in this water (Sontheimer, 1979; Kruithof *et al.*, 1991; Schippers, 1993; Degrémont, 1994a-c; Schippers and Kruithof, 1997).

---

[1] In this study the term "conventional water treatment processes" refers to all water treatment processes except those involving the application of ozone, activated carbon and/or membrane filtration.

This discussion, especially the part on pesticide removal, is currently most intense in the countries of the European Union (EU). This is not because of exceptionally poor quality of the raw water in these countries, but rather because their standards are more stringent than those promulgated by the World Health Organization (WHO), or those applied in other developed countries such as the USA and Japan. Based on the principle that pesticides must not be present in drinking water, the EU set the standard for any pesticide at 0.1 µg/l and the standard for the sum of all pesticides at 0.5 µg/l (ECC, 1980). In contrast, the WHO and the United States Environmental Protection Agency (USEPA) apply standards based on the acceptance of a certain health risk and, usually, allow much higher pesticide concentrations. For instance, they allow 2 µg/l and 3 µg/l of atrazine, respectively (WHO, 1993; Pontius, 1995).

1.1.2 Disinfection

The findings of Rook in 1974 initiated wide-scale research on chlorination by-products formation and control (Kruithof, 1986). The list of compounds eventually identified was long and frightening: besides trihalomethanes there was a whole host of other compounds including haloaceticacids, haloacetonitrils, halogenated ketones, chloropicrin, chloralhydrat, chlorophenols and MX (Kruithof *et al.*, 1991). Furthermore, the alternatives for chlorine, chlorine-dioxide and chloramine, were also found to result in by-products suspected to pose health hazards.

As a result, by the mid-seventies many water supply companies in the EU were considering restriction (or complete abandonment) of the use of chlorine and the application of ozone as the main disinfectant. For instance, Amsterdam Water Supply (AWS) started in 1970 with the reduction of the chlorine dose and completely abandoned chlorination in 1983. Ozone is highly effective as a disinfectant; however, the ozonation of water containing bromide results in the formation of bromate (Haag and Hoigné, 1983). This compound was shown to be possibly carcinogenic for humans (Kurokawa *et al.*, 1990). As a result, the WHO set a provisional guideline for bromate at 25 µg/l (WHO, 1993), while the proposed standard for the EU is 10 µg/l (EU, 1997). In the Netherlands, the National Institute of Public Health and Environmental Hygiene is currently considering 0.5 µg/l as the standard for bromate, if ozone is applied for the removal of organic micropollutants, while a higher value of 5 µg/l may be allowed when ozonation is used for disinfection (Van Dijk - Looijaard, 1994). At many water supply companies, bromate formation is expected to limit the ozone dose that may be applied.

The problem of adequate disinfection is further complicated by recent outbreaks of infectious diseases caused by the presence of *Giardia* cysts and *Cryptosporidium* oocysts in the drinking water prepared from surface water (Craun, 1988; Lisle and Rose, 1995). These outbreaks were caused by drinking water with a microbiological quality within the commonly applied standard, based on the absence of coliform bacteria in a certain volume of water sampled. Therefore, additional standards may be needed. However, if these standards are based on the absence of *Cryptosporidium*, *Giardia* or viruses, they require a sampling of extremely big volumes of water (Table 1.1). Due to such unfeasibly large sample volumes required, the USEPA promulgated the Surface Water Treatment Rule, and recently proposed its enhancement (Von Huben, 1991). These two rules prescribe treatment processes (and their performances) considered adequate for satisfactory disinfection, rather than prescribing the absence of indicator organisms in a certain volume of sampled water.

Table 1.1 Concentrations of various microorganisms that result in the commonly accepted risk of less than one infection per 10000 people per year (2 l per capita per day water consumption assumed).

Microorganism	Concentration in drinking water	Reference
Rotavirus	$2.2 \cdot 10^{-7}$ /l or < 1 per 4500 m^3	Regli *et al.* (1991)
Poliovirus	$2.6 \cdot 10^{-7}$ /l or < 1 per 4000 m^3	idem
Echovirus	$6.9 \cdot 10^{-5}$ /l or < 1 per 15 m^3	idem
Giardia	$6.7 \cdot 10^{-6}$ /l or < 1 per 150 m^3	idem
Cryptosporidium	$3.3 \cdot 10^{-5}$ /l or < 1 per 30 m^3	DuPont *et al.* (1995)

Another question posed by the aforementioned epidemics is how to ensure the adequate inactivation of *Cryptosporidium*. Such inactivation is difficult to achieve because the oocysts of this organism, which are the resting stages of the protozoa, are quite resistant to all chlorine-based disinfectants. Furthermore, high ozone doses that in combination with adequate particle removal may solve this problem, are expected to result in an unacceptable formation of bromate. On the other hand, UV-disinfection seems able to inactivate *Cryptosporidium* oocysts at rather low doses (< 10 mJ/cm^2).

Due to the problems associated with the attainment of the levels of disinfection required, more recent research aims to balance the chemical and microbial risks of disinfection (Craun *et al.*, 1994). In addition, alternative disinfection processes are being developed. A potential solution, provided the membrane integrity required can be achieved and effectively monitored, is the application of ultrafiltration or microfiltration, with UV or chlorine-based post-disinfection as an additional barrier for bacteria and viruses (Degrémont, 1994a; Schippers and Kruithof, 1997).

1.1.3 Removal of pesticides and other organic micropollutants

The presence of hundreds of pesticides and other micropollutants of industrial origin in surface water, is well-documented nowadays. The total number of compounds present is difficult to estimate, because suitable analytical methods are not available for many of them. For example, although 225 different pesticides were considered possibly relevant for drinking water treatment in the Netherlands in 1990, a suitable analytical method was available for only 70 (31%) of them (Hopman *et al.*, 1990a). Nowadays, 150 of these 225 pesticides (67%) can be analyzed (Hopman, 1997).

The presence of pesticides and other micropollutants in surface water poses a problem because the conventional water treatment processes cannot adequately remove many of these compounds. This was clearly shown when Amsterdam Water Supply found pesticide bentazon in its drinking water at concentrations higher than the EU standard of 0.1 μg/l (Smeenk *et al.*, 1988). As concluded by Hart and Jones in their 1989 review of published data, the conventional water treatment processes are considered effective in removing only low solubility pesticides, such as the organochlorine ones, unless pesticides are complexed with humic material. This is further corroborated by the data shown in Table 1.2 (Foster *et al.*, 1991) and Table 1.3 (Degrémont, 1994b).

Table 1.2 Mean pesticide concentrations (μg/l) measured after various treatment stages (Foster *et al.*, 1991).

Treatment stage	atrazine	isoproturon	mecoprop	lindane
Coagulation works				
river water	0.27	0.18	0.07	0.02
reservoir storage	0.25	0.15	0.05	0.01
coagulation and sedimentation	0.24	<.01	0.03	0.01
rapid sand filtration	0.22	<.01	0.03	0.01
Slow sand filtration works				
river water	0.33	0.14	0.08	0.03
reservoir storage	0.22	0.27	0.10	0.02
slow sand filtration	0.24	0.24	0.03	0.01
post-disinfection	0.22	0.22	0.03	0.01

Table 1.3 Removal of various pesticides by coagulation, sedimentation and/or rapid filtration (Degrémont, 1994b).

Pesticide	Removal (C_0 = 100 μg/l)
methoxychlor, DDT	90-98%
lindane, aldrin, dieldrin, parathion, endosulfan	10-60%
atrazine	< 10%

In response to the inadequate removal of organic micropollutants by the conventional water treatment processes, water supply companies in the late seventies started using more efficient processes. These processes were based on chemical and biological oxidation of micropollutants, and/or their adsorption onto activated carbon (Sontheimer, 1979).

One example of chemical oxidation is chlorine-induced oxidation. It was shown efficient for some compounds, such as phenylamide herbicides and organophosphorus insecticides (Hart and Jones, 1989). However, previously discussed by-products of chlorination are expected to limit its applicability. Another possibility is ozone-induced oxidation. It involves two types of reactions: direct oxidation by molecular ozone and oxidation by OH radicals formed through decomposition of ozone in water (Hoigné, 1988). Depending on the compound, the rate of oxidation by molecular ozone ranges from virtually instantaneous to very slow. Thus, among the many compounds present in raw water, only some will be oxidized significantly by molecular ozone. In contrast to molecular ozone, OH radicals are rather nonselective oxidants. This allows them to oxidize certain organic compounds, for instance triazines, that cannot be oxidized by molecular ozone at a sufficient rate (Meijers *et al.*, 1993). Processes that promote formation of OH radicals, either by H_2O_2 addition or UV-radiation, have been termed Advanced Oxidation Processes. However, despite the efficiency of these two forms of ozone-induced oxidation, their applicability is expected to be seriously limited by the formation of undesirable by-products, notably bromate. On the other hand, a combination of UV-radiation and H_2O_2 addition results in the production of OH radicals without any bromate formation.

Adsorption onto activated carbon has been recognized as a suitable technique for the removal of pesticides (Pontius, 1995). Activated carbon can be added to water as a powder (PAC), or water can be passed through filters filled with activated carbon granules (GAC). One main parameter that determines the removal efficiency for a certain compound is its adsorbability with respect to activated carbon. In principle, the more polar a compound is, the less adsorbable it is. Consequently, the removal of more polar pesticides increases the required PAC dosages or the empty-bed-contact-times and/or the frequency of GAC regeneration in GAC filters. This increases the cost of activated carbon application. The cost is also increased by the presence of organic matter of natural origin in treated water. The adsorption of these compounds onto activated carbon hinders the adsorption of target compounds, such as pesticides and other organic micropollutants. This results in higher PAC doses required, or in shorter running times of GAC filters before the breakthrough of target compounds (Sontheimer *et al.*, 1988).

A combination of ozonation and GAC filtration was proposed in the seventies as beneficial, because it combines chemical oxidation by ozone, adsorption onto GAC, and biological oxidation in GAC filters of compounds made biodegradable by ozone-induced oxidation. This recommendation was supported by the results of the pilot- and full-scale testing at, for instance, the Mülheim plant in Germany (Sontheimer *et al.*, 1978) and the treatment plants of the Zurich Water Supply (Schalekamp, 1989). In these tests, ozonation was shown to improve the removal of organic matter in GAC filters significantly, measured both as UV-absorbency and Dissolved Organic Carbon content. An ozone dose of about 0.5 mg O_3 per mg of organic carbon was recommended as, typically, required to achieve a good effect (Sontheimer, 1979).

The latest development regarding the removal of pesticides and other organic micropollutants from surface water is the application of membrane processes. The interest in membrane filtration is growing due to the continuously decreasing cost of its application and the superior quality of water produced by this process (Schippers and Kruithof, 1997). In particular, reverse osmosis is highly effective. The Water Supply Company of North-Holland recently completed construction of an integrated membrane system for the treatment of surface water, which comprises ultrafiltration and reverse osmosis. The role of ultrafiltration is to provide pretreatment for reverse osmosis and to contribute essentially to disinfection. The role of reverse osmosis in this system is not limited to the removal of organic micropollutants, but it also provides desalination, softening and disinfection.

1.2 NEW INTEGRAL CONCEPT AT AMSTERDAM WATER SUPPLY

The existing water treatment scheme at the Leiduin treatment plant of Amsterdam Water Supply (AWS) comprises the following processes:

- intake of Rhine River water - coagulation - sedimentation - rapid sand filtration - transport (WRK I/II water) - dune infiltration - aeration - rapid sand filtration - ozonation - softening - GAC filtration - slow sand filtration.

Dune infiltration plays an essential role in the treatment concept at Leiduin by providing the storage capacity and the equalization of water quality. However, due to the high environmental value of the dune area (Fig. 1.1), the Dutch government does not encourage an increase in dune infiltration. Therefore, when AWS recently anticipated the need to increase the capacity of its Leiduin treatment plant from 70 to 83×10^6 m^3 per year, they developed two new integral concepts for the treatment of Rhine River water. Neither of these two concepts uses dune infiltration. Furthermore, in contrast to water treatment schemes typically applied in The Netherlands, they do not use storage to provide for the periods when the Rhine River is heavily polluted, and to improve the quality of water by self-purification and equalization. Thus, the concept chosen has to offer reliable barriers against micropollutants and microorganisms present in Rhine River water, both under normal and accidental conditions.

The reliability of these barriers is of paramount importance, because pesticides such as triazines (especially atrazine), phenyl-ureic herbicides (including isoproturon), bentazon and dikegulac have been detected in the Rhine River at concentrations as high as 5 µg/l (Hopman *et al.*, 1990a,b). Furthermore, *Giardia* cysts, *Cryptosporidium* oocysts and viruses were also observed in the Rhine River at concentrations up to 57, 32 and 38 per liter, respectively (Medema *et al.*, 1996; Van Olphen *et al.*, 1993).

Figure 1.1 Dune infiltration at Amsterdam Water Supply.

The concept chosen also has to lower the high salinity, in particular the high chloride concentration of 150-200 mg/l, and the hardness of the raw water abstracted from the Rhine River. Chloride needs to be removed because a large part of a distribution system consists of uncoated cast-iron pipes (≈ 700 km), corrosion of which can be significantly reduced by lowering the chloride concentration below 80 mg/l (Van Soest, 1994). Considering that AWS abandoned chlorination in 1983, the biological stability of the drinking water needs to be ensured as well. Besides the Assimilable Organic Carbon (AOC) guideline of 10 µg Ac-C/l, AWS also aims for a DOC level of 1 mg/l. The purpose of the later is to control the growth of *Aeromonads* in the distribution network. DOC concentrations below 1 mg/l are not desirable since DOC is expected to play a role in the inhibition of the corrosion of cast iron pipes and to retard precipitation of calcium carbonate in water heaters.

It was with all these factors in mind that AWS developed the two new concepts. They consist of the following processes:

- concept I: intake of Rhine River water - coagulation - sedimentation - rapid sand filtration - transport (WRK I/II water) - slow sand filtration - reverse osmosis

- concept II: intake of Rhine River water - coagulation - sedimentation - rapid sand filtration - transport (WRK I/II water) - ozonation - GAC filtration - slow sand filtration - reverse osmosis

Both concepts include conventional pretreatment of Rhine River water, slow sand filtration, and reverse osmosis. The difference between them is the combination of ozonation and GAC filtration in the second one. This combination has been termed Biological Activated Carbon (BAC) filtration (Miller and Rice, 1978). BAC filtration is investigated at AWS since 1972, and has been applied on a full scale in 1992.

The role of reverse osmosis in these concepts is to provide desalination, softening, additional disinfection, and the additional removal of nutrients and micropollutants.

Slow sand filtration has two roles. One is to provide the biological stability of water and to reduce the content of colloidal matter in it, which allows less frequent cleaning of reverse osmosis membranes. Another role of slow sand filtration is to provide additional disinfection, namely the removal of bacteria, *Giardia* cysts and *Cryptosporidium* oocysts.

The main role of GAC filtration is the removal of organic micropollutants, especially pesticides and halogenated organic compounds, and a reduction in the overall content of organic matter. The removal of biodegradable organic matter is also important. It increases the biological stability of the water produced, and is expected to lower both the frequency of cleaning of the slow sand filters and the rate of fouling of the reverse osmosis membranes.

One role of ozonation is to provide an essential contribution to the disinfection of water. Another role is to oxidize partly organic matter present in water, and to significantly improve the removal of this matter, notably pesticides and other organic micropollutants, by GAC filtration. As stated by Graveland (1994) and Graveland and Van der Hoek (1995): "A low dose ozonation (0.5-1.0 mg O_3/l) increases the concentration of Assimilable Organic Carbon to 100-150 μg Ac-C equivalents per liter. Because of this, the mass of bacteria, or the total bioactivity, or the capacity for biodegradation of organic matter adsorbed on GAC is strongly (ca ten times) increased. Due to strong biodegradation, DOC values in general –and contents of undesirable, toxic organic micropollutants in particular– can be lowered considerably and for a long period."

Such extensive pretreatment by the combined ozonation and GAC filtration reduces the concentration of pesticides and other organic micropollutants in the membrane concentrate of the reverse osmosis to a large extent, resulting in a diminished discharge of these compounds in the environment. However, the application of ozone should not result in an unacceptable formation of bromate. This aspect deserves special attention because, due to high bromide (150-200 μg/l) and low DOC (2 mg/l) concentration in Rhine River water, formation of bromate may be expected to limit the ozone doses that may be applied.

Due to improved removal of pesticides and other organic micropollutants in the GAC filters, these filters can be run for a longer time before the regeneration of GAC is required. The regeneration of GAC, which is the removal and chemical oxidation of compounds adsorbed onto GAC, restores the adsorption capacity of GAC. Thus, after being regenerated, GAC can be reapplied. However, the regeneration of GAC is rather costly. Consequently, significant savings can be realized if the running time of GAC filters between the two regenerations of GAC can be extended.

Savings offered by postponed regeneration of GAC. To estimate the range of savings that can potentially result from the delayed regeneration of GAC, the annual costs of regeneration are calculated for empty-bed-contact-times of 20, 30 and 40 minutes, in case of regeneration every 6, 12, 24 and 48 months (Table 1.4). The calculation is done for the annual production capacity of 13×10^6 m^3 (Q_{avg} = 1500 m^3/h), which is planned for the extension of the treatment capacity at the AWS' Leiduin plant. The cost of one GAC regeneration, including GAC make up, is estimated as DFL 650 (US$ 350) per m^3 of GAC. This is about 40% of the cost of virgin GAC applied at Leiduin (Prinsen Geerligs, 1995).

Table 1.4 Annual costs of GAC regeneration.

EBCT	GAC volume	Costs (DFL) for regeneration once every			
(min)	(m^3)	6 months	12 months	24 months	48 months
20	500	650,000	325,000	163,000	81,000
30	750	975,000	488,000	244,000	122,000
40	1000	1,300,000	650,000	325,000	163,000

As shown by Table 1.4, postponing of GAC regeneration results in significant savings. The potential savings makes it attractive to investigate the mechanisms that underlie the expected beneficial effect of ozonation on the performance of GAC filters, and to try to maximize the time that can be allowed between two GAC regenerations.

An estimate of the cost involved in the application of ozone is also provided. However, the costs of ozonation do not necessarily decrease the savings offered by the less frequent regeneration of GAC receiving ozonated influent. Namely, ozonation frequently has disinfection as the main treatment objective. In such cases, the savings brought by the improved performance of GAC filters is the clear gain, and is not reduced by the costs of ozonation.

Based on Dutch experience (Wouters *et al.*, 1993), the investment costs for ozonation equal approximately DFL 300,000 (US$ 160,000) per kg O_3/h installed capacity. These costs are expected to be a linear function of the capacity installed. Ozone capacity required for the water flow of $Q_{max} = 1.25 \times Q_{avg} = 1875$ m^3/h and the maximum ozone doses of 1, 1.5 and 2 g O_3/m^3, and the resulting investment costs and annuity, are given in Table 1.5.

Table 1.5 Investment costs of ozonation and resulting annuity.

Ozone dose (g O_3/m^3)	Ozone capacity required (kg O_3/h)	Investment costs (DFL)	Annuity [1] (DFL)
1.0	1.88	563,000	58,000
1.5	2.81	844,000	87,000
2.0	3.75	1,125,000	116,000

[1] for 15 years return period and 6% interest

The annual operation and maintenance (O&M) costs, and the total costs (O&M costs and the annuity) of ozonation are given in Table 1.6. O&M costs were calculated in the following way. The annual ozone requirement was calculated for the annual production capacity of 13×10^6 m^3 and the average ozone doses of 0.8, 1.2 and 1.6 g O_3/m^3. The annual costs of energy and oxygen were calculated as the product of the annual ozone requirement and the cost of DFL 3 per kg O_3. Such cost of energy and (liquid) oxygen has been estimated for the AWS' Leiduin plant, and for the production of ozone at concentration of about 7% ozone by weight (Prinsen Geerligs, 1995). The annual O&M costs were calculated assuming that the costs of energy and oxygen account for 75% of the O&M costs (Langlais *et al.*, 1991).

Table 1.6 Annual operation & maintenance and the total costs of ozonation.

Average ozone dose (g O_3/m^3)	Ozone requirement (kg O_3)	Energy and oxygen costs (DFL)	O&M costs (DFL)	Total costs (DFL)
0.8	10400	31,000	42,000	100,000
1.2	15600	47,000	63,000	150,000
1.6	20800	62,000	84,000	200,000

1.3 RESEARCH PROJECT OF IHE, NORIT NV, Kiwa NV and AWS

The performances of the two aforementioned integral concepts for the treatment of Rhine River water were tested and evaluated by Amsterdam Water Supply (Fig. 1.2). Particular attention was paid to the following aspects:

- removal of pesticides, metabolites and other organic micropollutants by combined ozonation and GAC filtration, and by reverse osmosis;
- disinfection by ozonation and by reverse osmosis;
- control of the fouling and scaling of reverse osmosis membranes.

Figure 1.2 Technological laboratory at the Amsterdam Water Supply's Leiduin plant.

The mechanisms that play a role were studied in the context of a research project conducted by IHE, NORIT NV, Kiwa NV and AWS. The research presented in this thesis was conducted within the framework of this project. Its focus is on Biological Activated Carbon (BAC) filtration, which is a combination of ozonation and Granular Activated Carbon filtration.

Ozonation is expected to substantially improve the removal of pesticides and other organic micropollutants by GAC filtration. The general goal of this research is to identify and understand mechanisms that underlie the effect of ozonation on the removal of pesticides by

GAC filtration. This knowledge allows one to judge whether this combination provides a sound barrier against pesticides and other organic micropollutants. In addition, it allows optimization of the underlying mechanisms and, as a result, reduction in the cost of pesticide removal.

Before the scope of this research was defined, the processes anticipated as playing a role in the removal of pesticides by BAC filtration were analyzed (via desk study) and preliminary experiments were conducted. The main purpose of these experiments was to verify the hypothesis that the removal of pesticides by GAC filtration is improved due to ozone-induced oxidation of Background Organic Matter present in pretreated Rhine River water. The results of these analyses and experiments, and the formulated research objectives, are presented in the following chapter.

REFERENCES

Baker, M.N. (1949). *The quest for pure water.* AWWA, New York.

Craun, G.F. (1988). Surface water supplies and health. *Journal AWWA,* 2:40-52.

Craun, G.F., R.J. Bull, R.M. Clark, J. Doull, W. Grabow, G.M. Marsh, D.A. Okun, S. Regli, M.D. Sobsey and J.M. Symons (1994). Balancing chemical and microbial risks of drinking water disinfection, Part I. Benefits and potential risks. *J. Water SRT - Aqua,* 4:192-199.

Degrémont (1994 a,b,c). A: Membrane separation techniques: ultrafiltration. B: Removal of micropollutants: the case of pesticides. C: Membrane separation techniques: reverse osmosis and nanofiltration. *Proc. Technical symposium on water treatment,* Amsterdam.

DuPont, H.L., C.L. Chappell, C.R. Sterling, P.C. Okhuysen, J.B. Rose and W. Jakubowski (1995). The infectivity of *Cryptosporidium parvum* in healthy volunteers. *N. Engl. J. Med.,* 13:855-859.

European Community Council (1980). Directive concerning the quality of water intended for human consumption, no. 80/778/EC. *Official Journal,* L229, 30 August 1980.

European Union (1997). Commission's proposal for a Council Directive concerning the quality of water intended for human consumption. *Official Journal,* C213, 15 July 1997.

Foster, D.M., A.J. Rachwal and S.L. White (1991). New treatment processes for pesticides and chlorinated organics control in drinking water. *Journal IWEM,* 5:466-476.

Graveland, A. (1994). Application of biological activated carbon filtration at Amsterdam Water Supply. *Water Supply,* 14:233-241.

Graveland, A. and J.P. van der Hoek (1995). Introduction of biological activated carbon filtration at Leiduin. *H_2O,* 19:573-579 (in Dutch).

Haag, W.R. and J. Hoigné (1983). Ozonation of bromide-containing waters: kinetics of formation of hypobromous acid and bromate. *Env.Sci.Tech.*, 17:261-267.

Hart, J. and J.H. Jones (1989). *Removal of pesticides from water: a literature survey.* Water Research Center Report UM 1005, WRc, Swindon.

Hoigné, J. (1988). The chemistry of ozone in water. In: Stucki, S. (ed.) *Process technologies for water treatment.* Plenum Publishing Corporation.

Hopman, R., C.G.E.M. van Beek, H.M.J. Janssen and L.M. Puijker (1990a). *Pesticides and the drinking water supply in the Netherlands.* Report 113. Kiwa NV, Nieuwegein.

Hopman, R., Th.H.M. Noij and J.C. Kruithof (1990b). Dikegulac in Dutch bank-filtered water. *H_2O*; 18:482-486 (in Dutch).

Hopman, R. (1997). Personal communication.

Huisman, L. (1990). *Slow sand filtration.* Lecture notes, IHE, Delft.

Kruithof, J.C. (1986). *Chlorination by-products: production and control.* American Water Works Association Research Foundation and Kiwa NV, Denver.

Kruithof, J.C., J.C. Schippers and J.C. van Dijk (1991). Preparation of drinking water from surface water in the nineties. *H_2O,* 17:468-475 (in Dutch).

Kurokawa, Y., A. Maerkawa, M. Taqkahashi and Y. Hayashi (1990). Toxicity and carcinogenicity of potassium bromate - a new renal carcinogen. *Env. Health Pers.*, 87:309-335.

Langlais, B., D.A. Reckhow and D.R. Brink (1991). *Ozone in water treatment,* Lewis Publishers and AWWA Res.Found., Denver, p. 63-79 and 177-207.

Lisle, J.T. and J. B. Rose (1995). Cryptosporidium contamination of water in the USA and UK: a mini review. *J. Water SRT - Aqua,* 2:103-117

Miller, G.W. and R.G. Rice (1978). European water treatment practices - the promise of Biological Activated Carbon. *Civil Engr. - ASCE,* 2:81-83.

Medema, G.J., H.A.M. Ketelaars and W. Hoogenboezem (1996). *Cryptosporidium en Giardia in the Rhine River and Meuse River.* RIVM & RIWA Report no. 289202015,

Meijers, R.T., A.J. van der Veer and J.C. Kruithof (1993). Degradation of pesticides by ozonation and advanced oxidation. *Water Supply,* 11:309-320.

Pontius, F.W. (1995). An update of the federal drinking water regulations. *Journal AWWA*, 2:48-58.

Prinsen Geerligs, W.L. (1995). Design and building of ozonation and GAC filtration at Amsterdam Water Supply. *H_2O*, 19:580-588 (in Dutch).

Regli, S., J.B. Rose, C.H. Haas and C.P. Gerba (1991). Modeling the risk from Giardia and viruses in drinking water. *Journal AWWA*, 11:76-84.

Rook, J.J. (1974). Formation of haloforms during chlorination of natural water. *Water Treatment Exam*, 23:234-245.

Schalekamp, M. (1989). Future development of technology in the field of ozone in Switzerland. *Proc. Wasser Berlin*, 1.1.1-1.1.12.

Schippers, J.C. (1993). *Drinking water in turbulent flow.* IHE, Delft.

Schippers J.C. and J.C. Kruithof (1997). Membrane filtration in the next 10 to 25 years (treatment of surface water). *H_2O*, 6:179-182 (in Dutch).

Smeenk, J.G.M.M., O.I. Snoek and R.C. Lindhout (1988). Bentazon in the Rhine River-, rain- and drinking-water. *H_2O*, 7:183-185 (in Dutch).

Sontheimer, H., E. Heilker, M. Jekel, H. Nolte and F. Vollmer (1978). The Mülheim process. *Journal AWWA*, 7:393-396.

Sontheimer, H. (1979). Applying oxidation and adsorption techniques: a summary of progress. *Journal AWWA*, 11:612-617

Sontheimer, H., J.C. Crittenden and R.S. Summers (1988). *Activated carbon adsorption for water treatment*, AWWA - DVGW Forschungsstelle Engler Bunte Institut, Karlsruhe.

Van Dijk - Looijaard, A.M. (1994). Bromate. *Kiwa's Information Bulletin*, September 1994.

Van Olphen, M., E.van de Baan and A.H. Havelaar (1993). Removal of viruses by bank-filtration. *H_2O*, 3:63-66 (in Dutch).

Van Soest, E.A.M. (1994). Personal communication.

Wouters, J.W., W.F.J.M. Nooijen and P.J. de Moel (1993). What does the clean water cost? H_2O, 26:13:344-348 (in Dutch).

Von Huben, H. *(1991). Surface water treatment: the new rules.* AWWA, Denver.

WHO (1993). *Guidelines for drinking water quality.* WHO, Geneva.

Chapter 2

Ozonation and Atrazine Removal by Granular Activated Carbon Filtration: Process Analysis and Research Scope [1]

ABSTRACT—Ozonation is expected to improve the removal of pesticides by GAC filtration not only via the well-known effect of oxidation of pesticides, but also due to oxidation of Background Organic Matter (BOM) present in filter influent. Namely, an important part of BOM compounds will be partially oxidized because of ozonation. This will increase their biodegradability, and lower their adsorbability and molecular mass. Thus, enhanced biodegradation and reduced adsorption of BOM are expected in filters receiving ozonated rather than non-ozonated influent. Both biodegradation and adsorption of pesticides can be improved because of this.

Pilot plant experiments with pretreated Rhine River water spiked with about 3 μg/l of atrazine showed that ozonation results in incomplete oxidation of atrazine: about 25%, 45% and 65% of atrazine is oxidized when ozone doses of 0.5 mg/l, 1.0 mg/l and 1.5 mg/l are applied. For ozone doses up to 4 mg O_3/l, desethylatrazine and desisopropylatrazine (the two by-products monitored) are formed at concentrations of up to 20% and 5% of the initial atrazine concentration, respectively. By-products of atrazine oxidation are expected to be more biodegradable but less adsorbable than atrazine. This makes uncertain whether they are removed in GAC filters better or worse than atrazine.

Improved atrazine removal was observed in the pilot plant GAC filter that received ozonated (0.8 mg O_3/l) rather than non-ozonated pretreated Rhine River water. This improvement is thought to be due to ozone-induced oxidation of BOM, because atrazine was spiked after complete depletion of ozone and its concentration was the same (2.2±0.2 μg/l) in both ozonated and non-ozonated influent.

Research is defined to verify which of the anticipated processes and relationships underlie the improved atrazine removal observed in filters receiving ozonated influent.

---

[1] Part of this chapter has been published by E. Orlandini, M.A. Siebel, A. Graveland and J.C. Schippers (1994) in *Water Supply,* 14:99-108.

2.1 INTRODUCTION

Ozonation has a multi-functional role in the treatment philosophy of Amsterdam Water Supply (AWS), namely: disinfection, oxidation of pesticides and other organic micropollutants, and the improvement of the removal of these compounds by GAC filtration (Graveland, 1994). In this chapter, the two last mentioned roles are discussed. The pesticide atrazine was chosen as a model compound. A model compound needed to be chosen because detailed investigations for many organic micropollutants identified in Rhine River water were not possible. Such investigations would be too costly and, in addition, the appropriate analytical methods were not available for many of these compounds. Atrazine was chosen because –at the start of this study in 1992– its removal was relevant for AWS: atrazine was detected in pretreated Rhine River water at concentrations higher than 0.1 µg/l, which is the standard set by the European Union for any pesticide (Smeenk *et al.*, 1990). Moreover, atrazine was judged to be reasonable resistant to biodegradation. Its enhanced biodegradation in GAC filters receiving ozonated rather than non-ozonated influent would therefore, if verified, indicate enhanced biodegradation of other pesticides as well. Last but not the least, the analytical method needed to measure atrazine at concentrations below 0.1 µg/l was available.

Ozonation is expected to improve the removal of atrazine by GAC filtration. There are two reasons for this. First, a fraction of atrazine present in filter influent will be oxidized during ozonation. The resulting lower atrazine concentration will delay its breakthrough from GAC filters. This is expected from the theory of adsorption, and has been shown experimentally (Foster *et al.*, 1992; Degrémont, 1994). Only partial oxidation of atrazine is expected. Thus, the concentration of by-products of atrazine oxidation in the influent will be higher after ozonation. Considering that the by-products of atrazine oxidation are expected to be more biodegradable but less adsorbable than atrazine, it is difficult to predict whether they are removed in GAC filters better or worse than atrazine. Secondly, a fraction of Background Organic Matter (BOM) in the influent will be partially oxidized as well. Oxidation will increase the biodegradability, and reduce the adsorbability and molecular mass of these compounds. Thus, enhanced biodegradation[2] and reduced adsorption of BOM are expected in filters receiving ozonated rather than non-ozonated influent. Although it has not yet been clearly demonstrated, both biodegradation and adsorption of atrazine may be improved because of this.

---

[2] In this thesis the term "biodegradation" is used to describe biological activity which results in transformation of organic compounds and/or their complete breakdown to inorganic elements and compounds (mineralization).

The objectives of the work presented in this chapter are:
- to analyze, via desk study, processes anticipated as playing a role in the removal of atrazine by combined ozonation and GAC filtration;
- to relate oxidation of atrazine and formation of desethylatrazine and desisopropylatrazine to the ozone dose applied, and to determine GAC's adsorption capacity for these compounds;
- to verify the hypothesis that the removal of atrazine by GAC filtration is improved due to ozone induced oxidation of Background Organic Matter present in filter influent and to give an indication for the potential reduction in the annual cost of GAC regeneration that may result from it;
- to define, based on the results of this analysis and these experiments, the scope of the research to be done within a framework of this thesis.

The need for the aforementioned preliminary experiments came from the topic chosen for this research (*i.e.* to identify the mechanisms underlying the enhanced removal of pesticides in GAC filters receiving ozonated rather than non-ozonated water). If it turned out that the enhanced removal of pesticides in GAC filters receiving ozonated rather than non-ozonated influent does not occur, it would be necessary to change the topic of this research.

2.2 PROCESS ANALYSIS

Organic matter can be removed from water passing through GAC filters by two principal processes. They are adsorption onto GAC and biodegradation by bacteria present in these filters. Other processes that may occur, for instance oxidation catalyzed by activated carbon, are expected to play only a minor role.

2.2.1 Properties of granular activated carbon

Activated carbon is obtained by activation of suitable carbonaceous material (*e.g.* bituminous coal, lignite, peat, wood or coconut shells) which results in a material with a defined pore structure. The excellent adsorption capacity of (granular) activated carbon can be attributed to its high internal surface area ranging from 500 m^2/g to 1500 m^2/g (Boere, 1992) and the nature of its surface. The densities of GAC in backwashed and stratified filter beds range from 175 g/l for wood-based GACs to ca. 480 g/l for coconut-based GACs (Orlandini, 1992).

All activated (or reactivated) carbon types that are commercially available contain a certain quantity (circa 0,2% weight) of calcium in the form of CaO. Release of $Ca(OH)_2$ at the start of GAC filter operation may cause an increase in pH of filtered water to values as high as 11 (AWS, 1995). If filtered water contains sufficiently high HCO_3^- concentrations, the increased pH may cause precipitation of $CaCO_3$ in GAC pores resulting in their blocking. Precipitation of calcium carbonate can be prevented by addition of an acid to GAC filter influent which will lower its pH. Based on the experience of Amsterdam Water Supply, pH of GAC filter effluent is expected to equal the influent pH after approximately one week of filter running time.

Another characteristic of activated carbon which needs to be accounted for when starting up (pilot-plant) GAC filters is the initial chemisorption of oxygen. This phenomena has been widely observed and is so pronounced that GAC filters at the start of their operation produce anaerobic effluent. After about one week of filter running time, oxygen concentration in filter effluent approaches the concentration in filter influent (AWS, 1995).

Finally, before GAC filters are put in operation they need to be adequately backwashed in order to remove small particles (which may otherwise appear in filter effluent) and to ensure proper stratification of GAC bed which limits mixing of GAC during subsequent backwashes.

2.2.2 Adsorption in GAC filters

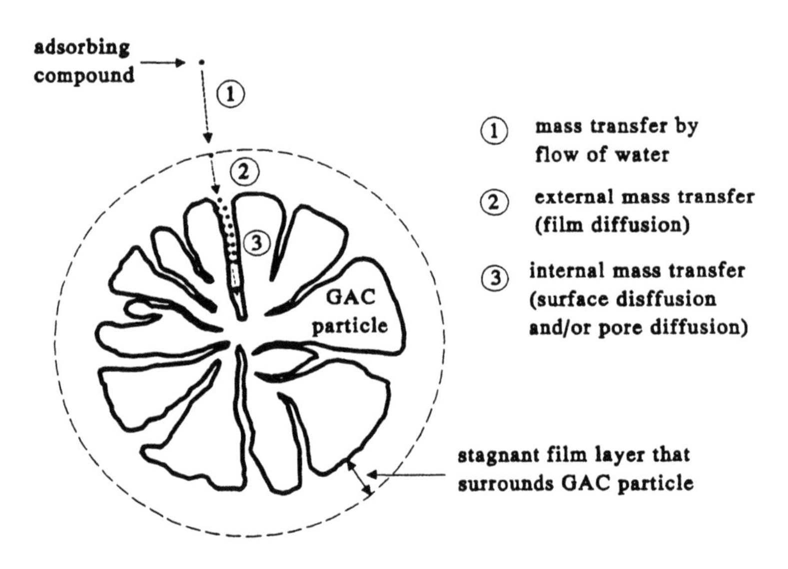

Figure 2.1 Mass transfer of a compound onto and into a GAC particle.

Adsorption capacity of GAC. The extent of the adsorption of a specific compound onto GAC is determined by the adsorption capacity of GAC for it. This capacity is the mass of the compound that can be adsorbed per unit mass of GAC (q) at equilibrium with a certain concentration of the compound in water (C). Adsorption capacity depends on GAC surface area available for adsorption, adsorbability of a compound with respect to activated carbon (which is related to the free Gibbs-energy of the carbon-adsorbate bond and, as such, function of temperature) and the concentration of the compound in water. The relationship between q and C, at constant temperature, is called adsorption isotherm. Usually, this relationship can be expressed by a simple Freundlich equation, parameters of which (K, n) are determined from the experiments:

$$q = K\, C^{n} \tag{2.1}$$

Adsorption kinetics. To get adsorbed onto GAC, the compound has first to reach it. Once brought by the flow of water to the vicinity of GAC particle (Fig. 2.1, step 1), the compound first needs to diffuse through the stagnant film layer surrounding this particle (Fig. 2.1, step 2). This is termed external mass transfer. Diffusion through the film layer follows Fick's first law of diffusion. The driving force for the diffusion is the difference between the concentration of the compound in the bulk solution and close to GAC particle (C_s). The rate of external mass transfer per unit mass of GAC (dq/dt) equals the product of this concentration gradient, the film mass transfer coefficient of the compound (k_f) and the external surface area available per unit mass of activated carbon (a_s):

$$\frac{dq}{dt} = a_s\, k_f\, (C - C_s) \tag{2.2}$$

Due to the porosity of a GAC particle, the compound can diffuse further inside it. Such internal mass transfer may go either along the surface of GAC pores, or through the water in these pores (Fig. 2.1, step 3). However, when modeling adsorption of a single compound in GAC filters, practically the same breakthrough of the compound is obtained if only surface diffusion, only pore diffusion, or a combination of both is assumed for the compound's internal mass transfer (Sontheimer *et al.*, 1988). Both modes of internal mass transfer are commonly applied. However, for the sake of simplicity, only the surface diffusion equation is given here. The rate of the internal mass transfer of a compound ($\delta q/\delta t$) is determined by the gradient of its loading onto GAC (q) over the radius of GAC particle (r) and its surface diffusion coefficient (D_s):

$$\frac{\partial q}{\partial t} = D_s \left(\frac{\partial^2 q}{\partial r^2} + \frac{2}{r}\frac{\partial q}{\partial r}\right) \tag{2.3}$$

The rate at which the compound is removed from water passing through GAC filters is limited by the rate of its external and/or internal mass transfer (Snoeyink, 1990). Because of this limitation, the compound adsorbs over a certain length of GAC bed. The part of the bed in which compound adsorbs at a given point of time is termed the mass-transfer-zone.

Background Organic Matter. In drinking water treatment, the influent of GAC filters usually contains many compounds. The compounds that need to be removed by GAC filtration are called target compounds. Compounds in the influent of GAC filters other than the target compounds have been termed either Natural Organic Matter (Sontheimer *et al.*, 1988) or Background Organic Matter (Najm *et al.*, 1991). Considering that many compounds present in Rhine River water are not of natural origin, the term Background Organic Matter (BOM) is used in this study. BOM adsorption in GAC filters interferes with the adsorption of the target compounds in two ways. They have been termed competitive adsorption of BOM and preloading of BOM.

Competitive adsorption of BOM. When BOM and target compound(s) simultaneously adsorb onto GAC, they compete for the adsorption sites available on activated carbon. Such competitive adsorption of BOM reduces the adsorption capacity of activated carbon for target compounds (*e.g.* atrazine). The decrease in adsorption capacity depends on the concentration and adsorbability of BOM. In principle, the higher the concentration of BOM and the more adsorbable the BOM compounds are, the lower the adsorption capacity for target compounds is. Due to competitive adsorption, previously adsorbed target compounds may also be desorbed (displaced) by more strongly adsorbing BOM compounds. Competitive adsorption of BOM is discussed further in Chapter 5.

Preloading of BOM. The preloading of BOM refers to the adsorption of BOM in GAC filters that occurs before the adsorption of target compounds such as atrazine. The part of the GAC filter bed in which BOM adsorbs in a given time is much larger and moves faster through the GAC bed than the part in which atrazine adsorbs. This is because, compared with atrazine, most of the BOM compounds are of lower diffusivity and are present at higher concentrations. As a result, an important part of BOM compounds adsorbs before atrazine in parts of GAC bed that are far from the inlet of the filter. The preloading of BOM has been identified rather recently, and was found to yield a pronounced decrease in both adsorption capacity of activated carbon for target compounds and the mass transfer rates of these compounds on and into GAC (Sontheimer *et al.*, 1988). The preloading of BOM is discussed further in Chapter 6.

Effect of temperature, water quality and contact-time. Temperature and water quality parameters of GAC filter influent, such as the concentration of target compound(s) and pH, are usually not constant. Their variation may affect adsorption in GAC filters. For instance, reduced influent concentration slows the mass transfer of target compounds, and may also cause desorption of previously adsorbed compounds because it shifts adsorption equilibrium. An increase in water temperature reduces the adsorption capacity of GAC, because adsorption is an exothermic process, but results in faster diffusion of compounds onto and into GAC. A change in pH that influences dissociation of weak organic acids and bases changes thereby their affinity for the adsorption onto activated carbon. Variation in temperature and pH does not have the same effect on all compounds and, therefore, it can change the effect of the competitive adsorption and preloading of BOM on the adsorption of target compounds. Thus, the parameters of an adsorption model need ideally to be set as a function of the water quality of filter influent. However, according to Weber (1972), temperature oscillations that are normally encountered in drinking water treatment are not expected to have a pronounced effect. Furthermore, atrazine adsorption isotherms determined at pH of 7, 8, 9, 10 and 11 all coincided (Knappe, 1996), while only a minor decrease in the extent of adsorption (from 28 to 26 mg DOC/g AC) was reported for Lake Constance humic substances (C_0 =3,9 mg DOC/l) for an increase in pH from 7.5 to 8.5 (Sontheimer *et al.*, 1988).

The minimum empty-bed-contact-time (EBCT) in a GAC filter can be determined by the requirement that there is no immediate breakthrough of target compound(s) in filter effluent. In order to achieve this, the filter bed needs to be deeper than the length of the mass-transfer-zone of target compounds. An increase in the bed depth, or EBCT, above the aforementioned minimum value raises the percentage of GAC in the filter for which the adsorption capacity is fully utilized when a certain breakthrough is reached. Full utilization means that the equilibrium adsorption capacity with regard to the concentration of target compounds in filter influent is reached. For an infinitely long EBCT, the percentage of GAC in the filter for which the adsorption capacity is fully utilized when a certain breakthrough is reached approaches 100%. This increase in the utilization of GAC adsorption capacity with longer EBCT is not linear and above a certain EBCT the benefit from the better utilization of adsorption capacity is lower than the costs involved in EBCT extension (more GAC, larger filters). The economically optimal EBCT (the one resulting in the lowest costs) is normally determined from the results of pilot-plant tests.

2.2.3 Biodegradation in GAC filters

The above discussion is related to the process of adsorption only. However, GAC filters are not sterile. When GAC is applied for the treatment of drinking water, nonhomogeneous bacterial coverage of GAC particles may be expected. This was clearly revealed, for instance, by electron scan microscopy of GAC particles exposed to 5 ppm solution of humic acids for 10 days (Weber *et al.*, 1978). Microorganisms, mostly rod-like bacteria of about 1.3 μm in length but also some protozoa's, were found scattered across the outer surface of GAC particles. The same may be seen in Figure 2.2, which shows electron-scan-micrographs of GAC particles taken from the filters operated at the AWS' Weesperkarspel plant (Boere and De Jonge, 1996). Therefore, besides adsorption, biodegradation of organic compounds also takes place in GAC filters.

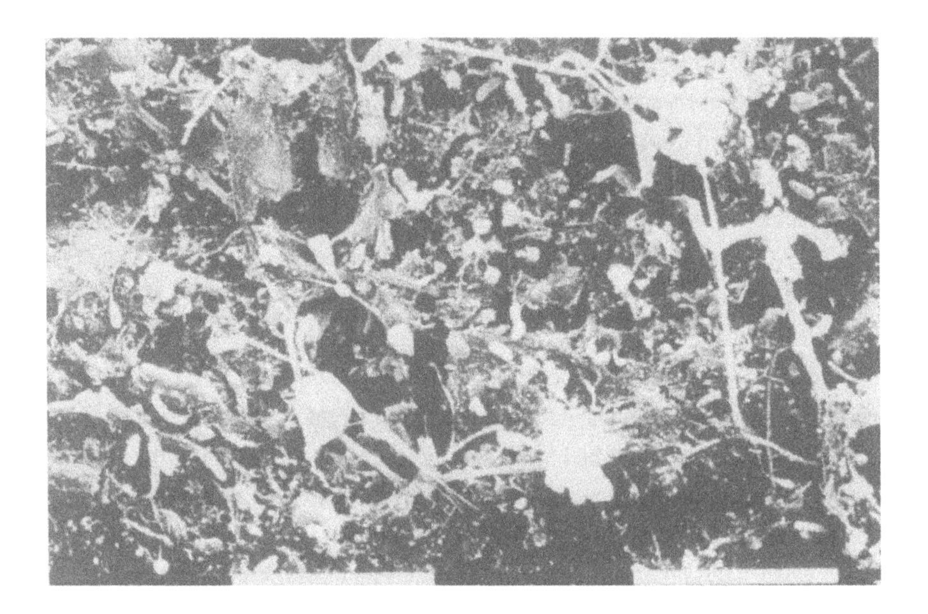

Figure 2.2 Electron scan micrograph of a GAC particle taken from filter receiving ozonated influent at the AWS' Weesperkarspel treatment plant.

Biodegradation of target compounds and BOM. In this study, a differentiation is made between the biodegradation of target compounds such as atrazine –removal of which is intended– and the biodegradation of BOM. A potential biodegradation of atrazine in GAC filters reduces the amount that needs to be adsorbed onto GAC and, as a result, delays the breakthrough of atrazine from these filters. One mechanism for atrazine biodegradation in GAC filters is catabolism, in which bacteria use atrazine as the main energy and carbon source. Biodegradation of atrazine in drinking water treatment via catabolism may be limited by the presence of atrazine at concentrations that are lower than the minimum concentration required to sustain the growth of bacteria, or by its only short-term presence at concentrations above these. In that situation, bacteria able to degrade atrazine might not be present at sufficiently high densities. The existence of the minimum compound's concentration below which bacteria cannot utilize the compound as a sole substrate was confirmed (for acetate) by McCarty *et al.*

(1981), both theoretically and experimentally. However, biodegradation of BOM in GAC filters may enhance biodegradation of atrazine. This is because bacteria grown on biodegradable organic matter may be able to metabolize atrazine as an additional source of carbon and/or nitrogen. The ability of biofilms grown on natural humic materials (*i.e.* peat, fulvic acid or ozonated lake water) to degrade target compounds such as 2-methylisoborneol, geosmin, phenol and naphthalene present at concentrations from 1 µg/l to 100 µg/l was shown in the experiments of Namkung and Rittman (1987) and DeWaters and DiGiano (1990). This mechanism is termed secondary utilization or co-metabolism.

Besides in enhanced biodegradation of target compounds such as atrazine, BOM biodegradation may also be expected to result in improved adsorption of the target compounds in GAC filters. This is because the removal of BOM via biodegradation rather than adsorption may reduce both the competitive adsorption and the preloading of BOM. The role that biodegradation of atrazine and BOM plays in the removal of atrazine in GAC filters is discussed further in Chapter 4.

Biodegradation of dissolved compounds. The discussion on the mechanism of biodegradation in GAC filters is ongoing (AWWA, 1981; Van der Kooij, 1983; Olmstead and Weber, 1991; Orlandini, 1992; Billen *et* al., 1992; Graveland and Van der Hoek, 1995; Graveland, 1995). In this discussion, biodegradation of compounds that are not adsorbed but are dissolved in water has been generally accepted.

Ying and Weber (1979), Billen *et al.* (1992) and Willemse and Van Dijk (1994) have modeled the process by which bacteria attached at the surface of filter medium degrade compounds present in water. The difference between these three models is in the assumptions they take regarding the mass transfer of compounds from the bulk solution to the attached bacteria. The model of Billen *et al.* neglects the mass-transfer of compounds and assumes that the concentration of a compound available to bacteria (C_s) equals the concentration in bulk solution (C). The model of Ying and Weber takes into account that a compound must first diffuse through liquid film surrounding filter material (Eq. 2.2), while the model of Willemse and Van Dijk allows for both diffusion through this liquid film and diffusion through the biofilm if it forms on a particle of filter material.

All three models propose that bacteria utilize compounds at a rate described by the Michaelis-Menten kinetics (Eq. 2.4). Thus, the compound is utilized at a rate that depends on the concentration of bacteria (B), the concentration at which this compound is available to bacteria (C_s), the half-saturation constant for that compound (K_s), and the maximum utilization rate for that compound per unit of bacteria ($b_{max,T}$) which, in turn, is function of temperature:

$$\frac{dC}{dt} = B\frac{C_s}{K_s + C_s}b_{\max,T} \tag{2.4}$$

The rate at which bacteria are growing is obtained as the product of the utilization rate (*dC/dt*), and the growth yield of bacteria for the compound being utilized (*Y*). In addition, models define the mortality rate of bacteria, and the rates at which bacteria attach to filter material and detach from it. These three rates are related via experimentally determined coefficients to bacterial growth rate and/or the concentration of bacteria in the filter.

According to these models, the extent of biodegradation in a filter is increased by higher concentration of biodegradable compounds in filter influent and their higher biodegradability (expressed as higher b_{max} and lower K_s), as well as by longer contact time in a filter and use of filter material with high affinity for attachment and low affinity for detachment of bacteria.

Biodegradation of adsorbed compounds (bioregeneration). The mechanism by which bacteria can utilize compounds adsorbed onto GAC is presently not well defined. Compounds can be adsorbed onto the outer surface of a GAC particle, or inside GAC's pores. Depending on their diameter (d), these pores have been termed macropores ($d > 50$ nm), mesopores (2 nm $< d <$ 50 nm) and micropores ($d < 2$ nm). Most bacteria are in a form of rods, cocci and filaments that range from 500 nm to 1000 nm in diameter (Laskin and Lechevalier, 1974). Thus, bacteria are expected to be able to colonize only the outer surface of GAC particles and the large macropores. Notwithstanding the fact that only 0.005% of the total GAC surface is attributed to the surface of macropores and the GAC particle itself (Boere and De Jonge, 1996) utilization of adsorbed compounds by bacteria may play an important role.

The possible role that bacterial enzymes may play in bioregeneration of GAC is presently not well understood. Perrotti and Rodman (1974) hypothesized that, although the size of the bacteria is too large for them to migrate into the vast majority of GAC pores, some enzymes excreted by the bacteria could easily diffuse into these pores and react with the adsorbed substrate. Van der Kooij (1983) hypothesized that enzymes can easily be deactivated during their transport through GAC pores due to adsorption of one or more of their functional groups. Xiaoijan *et al.* (1991) estimated the average diameter of the simplest and smallest (monomeric) enzymes as 3-4 nm and hypothesized that even these enzymes are too big to enter GAC pores that account for more than 95% of GAC surface area. All these hypotheses have yet to be verified.

Bioregeneration mechanism according to which adsorbed organic compounds are biodegraded upon their desorption from GAC pores has been widely accepted (AWWA, 1981; Sontheimer *et al.*, 1988; Snoeyink, 1990). As noted previously, desorption of already adsorbed compounds may occur due to several causes. First, it may occur due to an increased water temperature that shifts adsorption equilibrium. Secondly, desorption may occur due to reduced concentration of these compounds in water passing through the GAC filter. Such reduced concentration may be caused by biodegradation of these compounds in GAC filters, or by normal (*e.g.* seasonal) fluctuations of water quality. Thirdly, previously adsorbed compounds may be displaced due to competitive adsorption of compounds that are more adsorbable with respect to GAC.

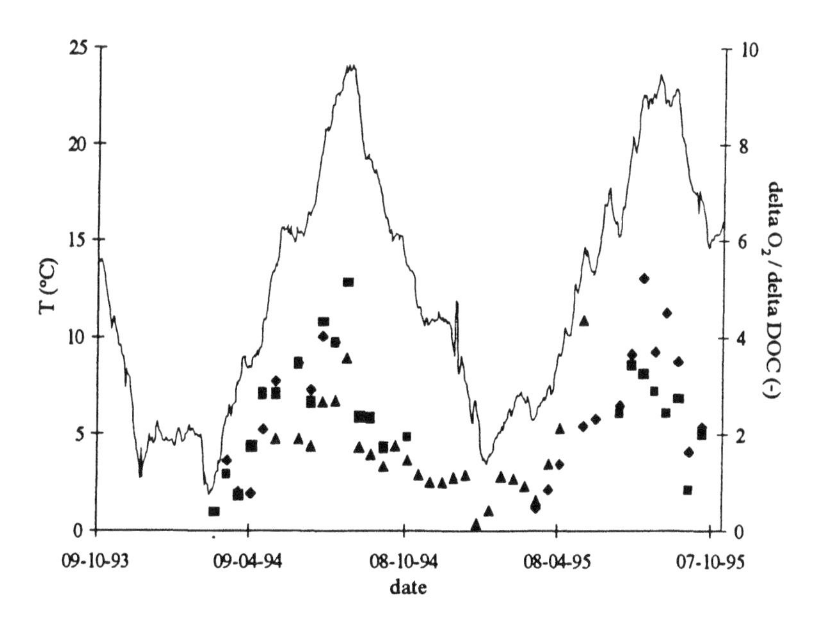

Figure 2.3 Water temperature (line) and the ratio between oxygen and DOC removal (points) in three full-scale GAC filters operated at the AWS' Weesperkarspel treatment plant ($\Delta O_2/\Delta DOC$ ratios are given only for filter running times longer than 200 days).

Biodegradation of previously adsorbed compounds, termed bioregeneration of GAC, has been first proposed based on the findings obtained in the late seventies at Bremen and Mülheim water works (Sontheimer *et al.*, 1988). These two studies showed that, after a prolonged operation time of GAC filters, the production of CO_2 and the consumption of oxygen in these filters are higher than can be explained by biodegradation of DOC being removed from the water being filtered. Such conclusion is based on the expected composition of organic matter as given by a formula $C_nH_{2n}O_n$, meaning that $\Delta O_2/\Delta DOC$ for the complete (to CO_2) biodegradation equals 2.7. Graveland and Van der Hoek (1995) postulated that bioregeneration of GAC plays a pronounced role. They observed at the AWS' Weesperkarspel plant an increased specific (per mass of DOC removed) oxygen consumption in GAC filters at higher water temperatures (Fig. 2.3).

Note that the actual extent of bioregeneration can be overestimated due to other phenomena which may play a role, such as increased removal of oxygen by physicochemical processes and/or increased endogenous respiration of bacteria at higher water temperature (AWWA, 1981). For instance, Van der Kooij and Visser (1976) observed an increase in oxygen consumption at higher water temperature in the experiments with both activated (NORIT ROW 0.8S) and non-activated carbon filters (in both filters ΔO_2 roughly doubled for a 10°C increase in water temperature). The influent of the GAC and NAC filter was filtered and aerated Rhine River water, so accumulation in these filters of suspended organic matter or bacteria from filter influent is not likely to play a dominant role. The increased oxygen consumption at higher water temperature was thought to be attributable to the slight dissimilation (about 30%) taking place during the periods of low water temperature and leading to an accumulation of cell material in the filters. The accumulation of cell material was expected to result in an inhaling effect (higher endogenous oxygen consumption) during the periods of high water temperatures, with the compounds accumulated in the biomass being degraded. Since the experiments lasted several months the effect of physicochemical oxygen consumption could not play a dominant role.

Effect of temperature, water quality and contact-time. An increase in water temperature increases both the rate of mass transfer of compounds and the rate of their biodegradation. Furthermore, an increased water temperature can result in desorption of previously adsorbed compounds because it shifts adsorption equilibrium. As noted previously, such desorption can enable degradation of these compounds by bacteria in GAC filters. For a certain type of bacteria there is normally an optimum pH range at which these bacteria degrade organic matter at the highest rate. A change in pH will lower the rate of degradation, while sufficiently high change of pH may completely inhibit degradation by these bacteria. Biodegradation of organic matter can also be limited by insufficient availability of essential nutrients and microelements. Similarly, biodegradation of organic compounds can be prevented –or shifted to another pathway– due to the lack of oxygen.

As described by the aforementioned models for biodegradation in GAC filters, prolonged contact-time in these filters increases the extent of biodegradation. However, measured as DOC removal, the extent of biodegradation per volume GAC decreases with the increasing bed depth or contact-time (Billen *et al.*, 1992; Orlandini, 1992). Considering that the prolonged contact-time increases the investment cost of GAC application, it is likely that an (economically) optimal contact-time can be defined. It has also been hypothesized that long contact times enable biodegradation of somewhat recalcitrant organic compounds that are not biodegraded when easily biodegradable organic matter is present at high concentrations (Graveland, 1994). This hypothesis, based on the mechanism of sequential utilization of organic matter (Brock and Modigan, 1988), is discussed in more detail in Chapter 4.

2.2.4 Effect of ozonation

The preceding section related to adsorption and biodegradation, the two principal processes by which organic matter can be removed in GAC filters. This section describes the effect of ozonation on the removal of atrazine in these filters. Two effects play a role. Namely, the well-known effect of ozone-induced oxidation of atrazine, and the effect of ozone-induced oxidation of Background Organic Matter present in filter influent.

Ozone-induced oxidation of atrazine. Atrazine oxidation by molecular ozone is, normally, not expected to be significant. This is because of the slow first order reaction between atrazine and molecular ozone, for which the rate constant is only 6 $M^{-1}s^{-1}$ (Yao and Haag, 1991). Only compounds for which this constant is higher than 100 $M^{-1}s^{-1}$ are expected to be oxidized significantly by molecular ozone. OH radicals formed by ozone decomposition are expected to contribute significantly to atrazine oxidation. However, they are also scavenged by other compounds present. Because of ozone-induced oxidation, atrazine concentration in the influent of GAC filters is reduced. This, in turn, delays the breakthrough of atrazine from these filters (Foster *et al.*, 1992; Degrémont, 1994).

Ozonation is not expected to result in complete oxidation (to CO_2) of atrazine. Thus, the concentration of atrazine oxidation by-products in the influent of GAC filters will be higher after ozonation. Desethylatrazine was shown to be the prevalent atrazine by-product (formed at concentrations of up to 20% of the initial atrazine concentration), while desisopropylatrazine, hydroxyatrazine, desisopropylatrazine amide and many other by-products were detected also (Legube *et al.*, 1987; Adams and Randtke, 1992). The degree to which these compounds will be formed and the efficiency at which they will be removed by GAC filtration is of interest. This is especially the case for desethylatrazine and desisopropylatrazine, because the Recommendations of the Netherlands Water Supply Association (VEWIN, 1993) treat these two compounds in the same way as atrazine. Consequently, their concentration in drinking water needs to be lower than 0.1 μg/l. In addition, they count toward the sum concentration of all pesticides that should not exceed 0.5 μg/l.

Ozone-induced oxidation of BOM. By ozone oxidized Background Organic Matter compounds are expected to be less adsorbable with respect to activated carbon than the parent compounds (Langlais *et al.*, 1991; Sontheimer *et al.*, 1988). At the same time, just a very small part of BOM will be oxidized to carbon dioxide. Based on these considerations only, the efficiency of BOM removal should be reduced in GAC filters receiving ozonated rather than non-ozonated influent. However, ozonated BOM compounds are also expected to have a higher diffusion coefficient (due to reduced molecular mass) than the parent compounds (Langlais *et al.*, 1991; Sontheimer *et al.*, 1988). Furthermore, ozonation is commonly found to increase the concentration of biodegradable organic matter in water (Van der Kooij *et al.*, 1989; Huck *et al.*, 1991). In GAC filters biodegradable organic matter is expected to be transferred to less biodegradable compounds and carbon dioxide. As noted before, the extent of biodegradation of organic matter depends, among other things, on the nature of the organic compounds, contact time and temperature.

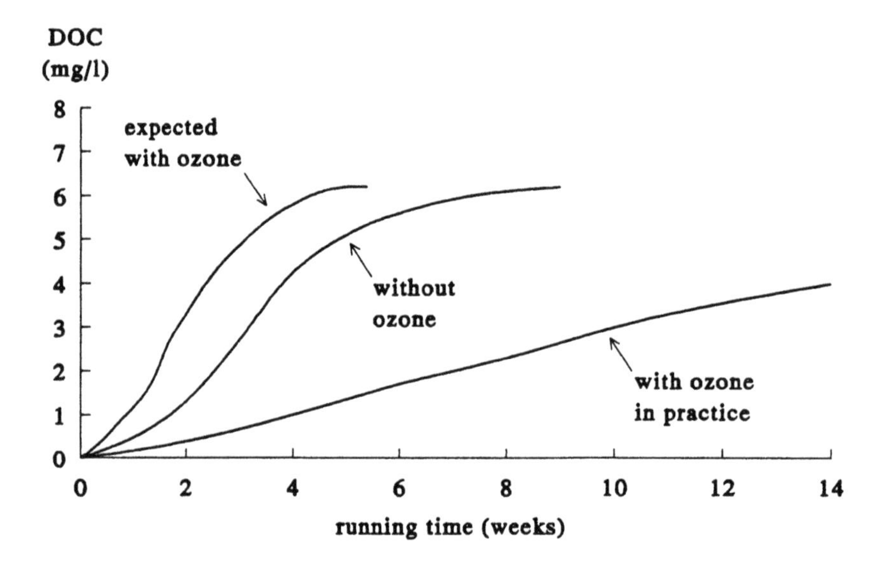

Figure 2.4 Effluent DOC concentrations observed in GAC filters receiving non-ozonated and ozonated influent at the AWS' Weesperkarspel treatment plant and those expected based on increased polarity of ozonated DOC-compounds (Graveland, 1994).

Graveland (1994) recorded at the AWS' Weesperkarspel treatment plant an improved removal of organic matter (measured as DOC concentration) in GAC filters receiving ozonated rather than non-ozonated influent (Fig. 2.4). Although an increase in water temperature reduces the adsorption capacity of GAC, DOC removal in filters receiving ozonated influent was higher during the periods of high water temperature (spring-summer) than during the periods of low water temperature (autumn-winter). Oxygen consumption and lowering of effluent pH in these filters were also increased by higher water temperature, indicating that the removal of organic matter in GAC filters can be attributed to its biodegradation. Graveland (1995) proposed that biodegradation during summer months is so fast and effective that the internal-surface of GAC is almost freed from adsorbed compounds (bioregeneration of GAC), which makes this surface available for the adsorption of organic compounds in autumn and winter. The spring increase in water temperature can than initiate biodegradation of organic compounds that were adsorbed rather than biodegraded during the periods of low water temperature. As noted previously, this mechanism is supported by the higher production of CO_2 and consumption of oxygen in GAC filters during the periods of high water temperature than can be explained by biodegradation of DOC being removed from water being filtered ($\Delta O_2/\Delta DOC$ higher than 2.7; see Fig. 2.3). As shown by Figure 2.5, removal of DOC and oxygen in the GAC filters at Weesperkarspel were found to occur at constant rate over the whole filter bed depth (empty-bed-contact-time of 40 minutes), indicating that biodegradation of organic matter in these filters is a pseudo zero-order process (Graveland, 1995).

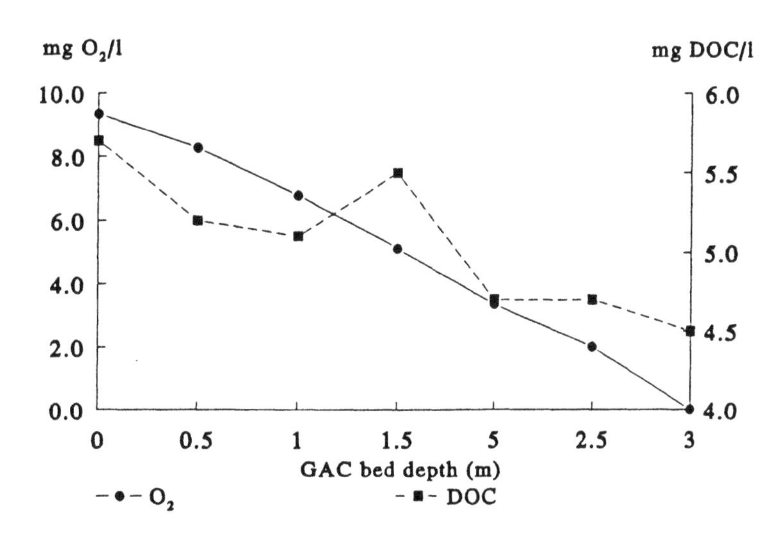

Figure 2.5 Oxygen and DOC concentration as a function of the filter bed depth in the full-scale GAC filter operated at the AWS' Weesperkarspel treatment plant (June 1997, filter running time 490 days).

Graveland (1995) proposed the mechanism for Biological Activated Carbon filtration (combined ozonation and GAC filtration) that contains the following steps:

- by ozone induced increase in AOC concentration of treated water and breaking of large humic-molecules into smaller, faster to diffuse blocks;
- transfer of reactants (oxygen, organic matter measured as DOC and AOC, pesticides and other organic micropollutants) from the bulk solution via the film layer surrounding GAC particle onto this particle;
- transport of the aforementioned reactants through the pores of activated carbon;

- adsorption of these reactants onto internal surface of GAC (in macro-, meso- and micropores);
- growth of microorganisms in GAC pores (due to high AOC concentration) and biodegradation of organic compounds to CO_2 and water (the increased density of microorganisms increases the capacity for biodegradation);
- desorption of the produced CO_2 from the internal surface of activated carbon;
- transport of CO_2 through the pores of activated carbon to the outer surface of GAC particle;
- transport of CO_2 through the stagnant film layer surrounding the GAC particle into bulk solution.

Graveland (1995) further concluded that the mass transport through the GAC pores is the slowest among the aforementioned steps and that this step, therefore, determines the overall rate of the BAC filtration process. Furthermore, he asserted that the increased biological activity in GAC filters receiving ozonated influent also enhances the biodegradation of the organic compounds which are not (partly)oxidized by ozone. Therefore, ozone-induced oxidation of BOM may be hypothesized to enhance the biodegradation of atrazine and other micropollutants in GAC filters, and/or to improve their adsorption in these filters.

This hypothesis is further supported by the following two considerations.

First, ozone-induced oxidation of BOM may be expected to stimulate biodegradation of atrazine in GAC filters because it increases the concentration of biodegradable organic matter in their influent. Due to such increased concentration of biodegradable organic matter, more bacteria are expected to colonize filters receiving ozonated influent than filters receiving non-ozonated influent. As mentioned previously, these bacteria may be able to use atrazine as an additional source of carbon and/or nitrogen. This mechanism is discussed further in Chapter 4.

Second, ozone-induced oxidation of BOM may also be expected to lower the competitive adsorption of BOM. This is because of reduced concentration of BOM that can be adsorbed in filters receiving ozonated rather than non-ozonated influent (a higher fraction of BOM is biodegraded) and because of the lower adsorbability of BOM compounds after ozonation. In addition, a larger part of the compounds which are not oxidized by ozone may be biodegraded to less adsorbable compounds. Due to these reasons ozonation may also be expected to reduce the preloading of BOM. In addition, the preloading of BOM can also be reduced due to decreased molecular mass of oxidized BOM compounds. Because of reduced molecular mass, oxidized BOM can result in less physical blocking of GAC pores. A pore can be blocked by a molecule of BOM if the pore becomes too narrow for BOM to diffuse further inside it. In this way, other adsorbing compounds are prevented from reaching the free adsorption sites that are available further inside the pore. Reduced competitive adsorption and preloading of BOM in filters receiving ozonated influent can result in higher adsorption capacity of GAC for target compounds such as atrazine, and in higher mass transfer rates of these compounds onto and into GAC. This, in turn, will delay the breakthrough of atrazine from these filters. The mechanisms of BOM competitive adsorption and its preloading are discussed further in Chapters 5 and 6, respectively.

The anticipated processes and relationships through which ozone-induced oxidation of BOM can improve removal of atrazine by GAC filtration are given in Figure 2.6. The expected higher extent of bioregeneration in GAC filters receiving ozonated rather than non-ozonated influent is thought to result from the enhanced biodegradation of both BOM and atrazine in filters receiving ozonated rather than non-ozonated influent.

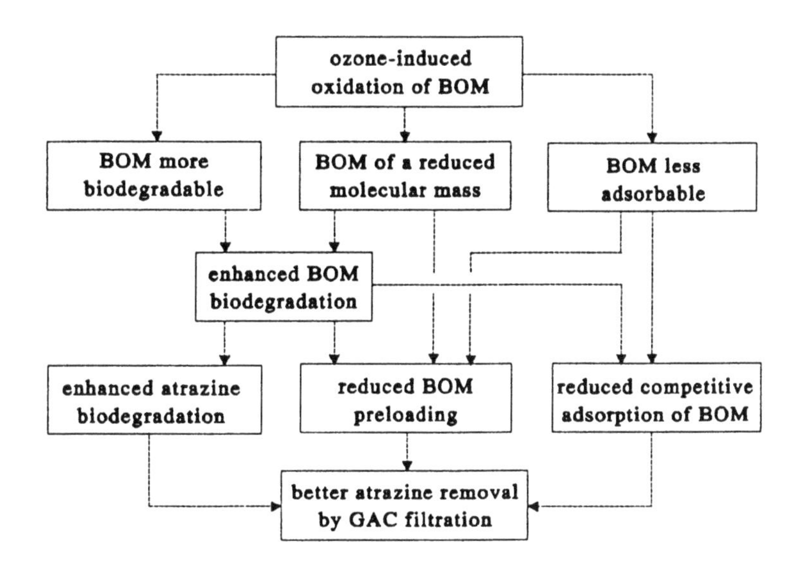

Figure 2.6 Processes and relationships anticipated to underlie the beneficial effect of ozone-induced oxidation of Background Organic Matter (BOM) on atrazine removal by GAC filtration.

2.3 MATERIALS AND METHODS

2.3.1 Atrazine oxidation

The Rhine River water pretreated with coagulation, sedimentation and rapid sand filtration (further termed pretreated Rhine River water) was ozonated in a counter-current column (height 5 m, diameter 0.1 m). This column was operated at the Amsterdam Water Supply's

Leiduin treatment plant. Ozone was produced from oxygen, at a gas-concentration of about 30 mg/l. The applied ozone doses ranged from 0.2 mg/l to 4 mg/l. The flow of water was either 1000 l/h for ozone doses up to 1 mg/l, or 500 l/h for higher doses. The corresponding detention time in the ozone-dosing column is 2.5 minutes and 5 minutes. The experiments were conducted twice. Water quality parameters of pretreated Rhine River water on these two occasions are given in Table 2.1.

Table 2.1 Water quality in experiments for atrazine oxidation.

	T (°C)	pH	DOC (mg/l)	$[HCO_3^-]$ (mg/l)	UV_{254nm} (1/m)
1st experiment	12	8.0	2.0	153	6.1
2nd experiment	7.5	8.0	2.1	162	6.4

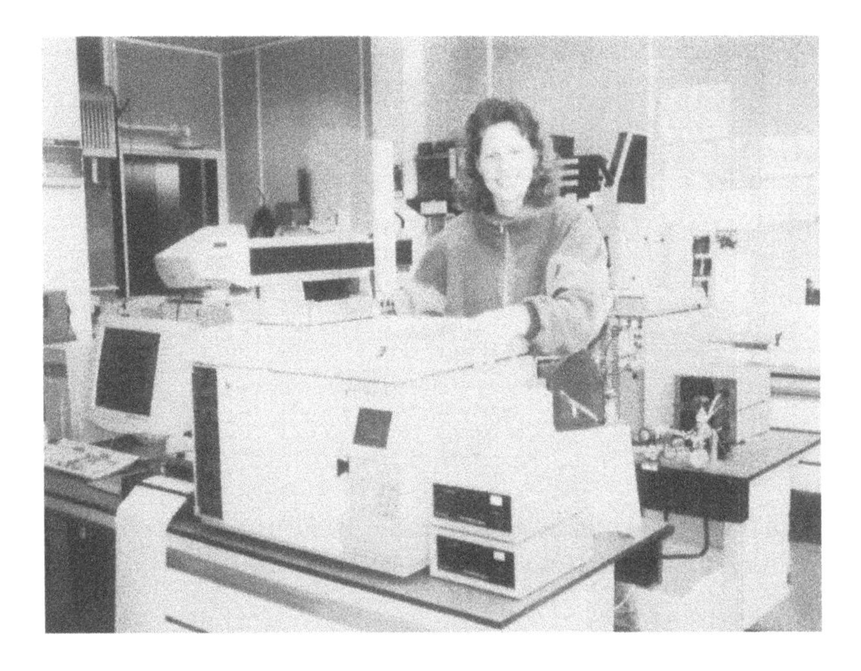

Figure 2.7 Ada Vooijs, who analyzed most of the atrazine samples collected in this study, and "her" Chrompack 2000.

Pretreated Rhine River water was spiked with 2.9-3.6 μg/l of atrazine before ozonation. Such a high concentration of atrazine was applied to ensure that the concentration of atrazine remaining after ozonation and the concentration of the two atrazine by-products monitored would be above the detection limit of the analytical method applied. Ozone was not quenched after sampling and, as a result, atrazine, desethylatrazine and desisopropylatrazine were measured after all ozone had reacted.

Atrazine and its two by-products were measured via an analytical method standardized by Amsterdam Water Supply (AWS, 1993). This analytical method involves liquid-liquid extraction with ethyl-acetate, and gas chromatography with nitrogen-phosphorous detection (Fig. 2.7). The concentration of atrazine is determined by comparing the ratio between the area of atrazine peak and the peak of an internal standard (phosphamidone) in a sample analyzed and in the standard with known atrazine concentration. The recovery of atrazine from pretreated Rhine River water is 93%. The detection limit of the method is 0.03 μg/l for all three compounds, while method accuracy is ± 5% at 1 μg/l of their concentration.

GAC's adsorption capacity. Adsorption capacity of GAC for atrazine, desethylatrazine and desisopropylatrazine was deduced by determining adsorption isotherms for these three compounds. These isotherms were determined for each compound separately. The GAC used was NORIT ROW 0.8S, obtained directly from the manufacturer (NORIT, 1993). GAC was pulverized in a way that 98% of particles pass the 75 μm sieve and, prior to its pulverization, was washed for one week by means of Soxhlet extraction using ultrapure water (*i.e.* water passed through resin, GAC and UF membrane filter). This was done to ensure that aluminum, calcium and other impurities are released from activated carbon prior to isotherm tests. All activated carbon fractions obtained were combined and suspended (0.3-5 g AC/l) in demineralized-deionized water with TOC concentration below the detection level of analyzer (0.05 mg/l). Various volumes (1-20 ml) of this suspension were added to bottles, which were then filled up to an end-volume of 1 liter with demineralized-deionized water containing 30 mg/l of a given compound. The concentration of activated carbon in these bottles ranged from 0.3 mg/l to 4.2 mg/l. Control bottles with no carbon addition were used to measure the initial concentration in the solution. In bottles with carbon addition, the concentration of atrazine, desethylatrazine or desisopropylatrazine remaining in the solution was measured after these bottles were stirred for 96 hours at 25°C. This time was found to be sufficient to reach the adsorption equilibrium (data not shown).

2.3.2 Effect of BOM oxidation

An ozonation - GAC filtration pilot plant was operated at the Amsterdam Water Supply's Leiduin treatment plant (Fig. 2.8). The influent of the pilot plant was Rhine River water, pretreated with coagulation, sedimentation and rapid sand filtration. Water quality parameters of this water are given in Table 2.2.

Figure 2.8 Ozonation - GAC filtration plant operated at the Amsterdam Water Supply's Leiduin plant.

Table 2.2 Water quality during pilot plant operation (average values and ranges).

DOC (mg/l)	UV_{254nm} (1/m)	T (°C)	pH
2.1 (1.6-3.2)	6.2 (5.2-7.7)	12.0 (5-23.5)	7.8 (7.4-8.3)

The pilot plant comprised one gravity GAC filter (d = 0.265 m) filtering non-ozonated and one filtering ozonated influent. The ozone dose was 0.8 mg/l. To achieve the same oxygen concentration in ozonated and non-ozonated influent, they were both passed through cascade aerators (four trays, the total height of 1 m) before they reached GAC filters. GAC filters were filled to a bed depth of 1.1 m with the extruded activated carbon ROW 0.8S manufactured by NORIT NV. Before the filters were put into operation, ultra-pure water (*i.e.* permeate of the reverse osmosis pilot plant) was percolated through them for 11 days (v = 1 m/h), after which they were backwashed to ensure the removal of fine material and the stratification of the GAC bed. This washing of GAC for 11 days is expected to eliminate potential increase of pH at the start of normal filter operation. Amsterdam Water Supply has namely observed that aluminum and calcium are released from virgin and regenerated activated carbon grains (AWS, 1995). Such release results in an increase in pH, which may cause precipitation of calcium carbonate and blocking of GAC pores. In addition the dissociation of BOM will be larger at high pH values, which will reduce its adsorbability.

GAC filters were operated with an empty-bed-contact-time (EBCT) of 20 minutes. Additional sampling points were provided at the EBCT of 7 minutes. Two monitoring points per GAC filter were chosen based on the available analytical capacity. The EBCT of 7 minutes was chosen because it allows clear breakthrough of atrazine (ca. 50%) in a relatively short period of time (ca. 6 months). This allows prompt testing of the hypothesis that ozonation improves removal of atrazine in GAC filters and, if required, prompt adjustment of the research scope. The EBCT of 20 minutes was chosen because it has been applied at several water treatment plants. Moreover, longer EBCTs (for instance 40 minutes) was expected to require several years of filter operation for clear breakthrough of atrazine. The influent of GAC filters was spiked with 2.2 ± 0.2 µg/l of atrazine. The concentration of atrazine was the same in both ozonated and non-ozonated influent, because atrazine was spiked to ozonated influent only after complete depletion of ozone (residual ozone concentration < 0.01 mg/l).

During the initial two years of pilot plant operation, atrazine, desethylatrazine and desisopropylatrazine concentrations were regularly (on average once in every four weeks) monitored in the influents of the GAC filters, and their effluents at a bed depth of 0.35 m and 1.1 m. The mass of atrazine removed (M_r) during this time in the top 0.35 m of the GAC filter bed and in a whole filter bed (depth of 1.1 m), was determined by summing the products of the water flow (Q), filter running time and the difference between the influent (C_0) and effluent concentration of atrazine:

$$M_r = \sum Q\,t\,(C_0 - C) \tag{2.5}$$

2.4 RESULTS AND DISCUSSION

2.4.1 Atrazine oxidation

Ozonation of pretreated Rhine River water was found to result in incomplete oxidation of atrazine (Fig. 2.9). About 75%, 55% and 35% of atrazine remained in pretreated Rhine River water after an ozone dose of 0.5 mg/l, 1 mg/l and 1.5 mg/l had been applied. Note that the oxidation of atrazine was allowed until the complete depletion of ozone, while in the full-scale plant GAC filtration will quench ozone after about 30 minutes of retention time. Thus, the actual extent of atrazine oxidation is expected to be lower than shown by Figure 2.9. The maximum atrazine concentration observed in pretreated Rhine River water in the period 1990-1995 was 0.3 µg/l. Consequently, for ozone doses up to 1.5 mg/l, the concentration of atrazine after ozonation will still be higher than 0.1 µg/l, and atrazine has to be removed in GAC filters.

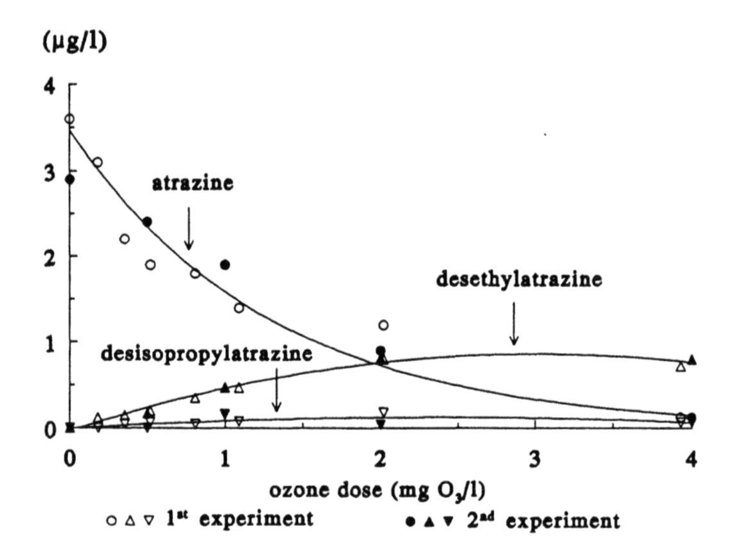

Figure 2.9 Oxidation of atrazine and formation of its by-products desethylatrazine and desisopropylatrazine as a function of the ozone dose.

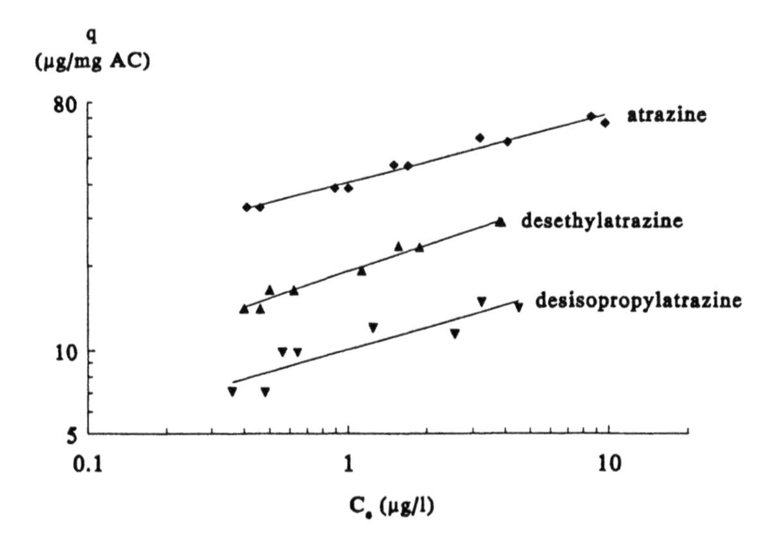

Figure 2.10 Single-solute isotherms for atrazine and its by-products desethylatrazine and desisopropylatrazine.

Ozonation of atrazine-spiked pretreated Rhine River water was found to result in the formation of both desethylatrazine and desisopropylatrazine. For ozone doses up to 4 mg O_3/l, they were formed at concentrations of up to 20% and 5% of the initial atrazine concentration, respectively. Thus, for the aforementioned maximum initial atrazine concentration of 0.3 μg/l, less than 0.1 μg/l of both desethylatrazine and desisopropylatrazine is expected to be formed. Most likely, oxidation of atrazine resulted in the formation of more by-products than just the two measured in this study; for instance, hydroxyatrazine, desisopropylatrazine amide, and other compounds identified by Legube *et* al. (1987) and Adams and Randtke (1992). The individual concentrations of these compounds are expected to be below the concentration of desethylatrazine, which was the main by-product of atrazine oxidation in these two studies.

GAC's adsorption capacity. Being by-products of atrazine oxidation, desethylatrazine and desisopropylatrazine are expected to be more biodegradable and less adsorbable than atrazine. Their higher biodegradability was shown in studies conducted by Kearney *et al.* (1988), while their lower adsorbability with respect to activated carbon is shown in this study (Fig. 2.10). These characteristics make it difficult to predict whether or not the oxidation products of atrazine will be better removed in GAC filters than atrazine. However, due to the limited extent to which they are formed by ozonation, the need for their removal may be questioned.

2.4.2 Effect of BOM oxidation

During the two years of pilot plant operation, the performance of the GAC filter that received ozonated influent was clearly better than that of the GAC filter that received non-ozonated influent (Fig. 2.11). The mass of atrazine removed in GAC filters was calculated by the equation 2.5. The mass of atrazine removed in the top 0.35 m (EBCT 7 min) and in the whole 1.1 m (EBCT 20 min) of the filters was 3.5 g and 7.4 g of atrazine in the filter receiving non-ozonated influent and 4.8 g and 8.9 g of atrazine in the filter receiving ozonated influent. Thus, in the top 0.35 m of the GAC bed about 40% more atrazine was removed from ozonated rather than non-ozonated influent, while this difference was 20% for the whole bed depth.

No pronounced effect of temperature on the removal of atrazine in either GAC filter was observed. Therefore, it has to be concluded that bioregeneration in GAC filters has not been demonstrated in this experiment.

Considering that atrazine was spiked to ozonated influent only after complete depletion of ozone, the results obtained clearly show that ozonation improves the removal of atrazine by GAC filtration not only via the well-known effect of ozone-induced oxidation of atrazine, but also due to ozone-induced oxidation of Background Organic Matter (BOM) present in filter influent. The resulting delayed breakthrough of atrazine in GAC filters receiving ozonated influent allows less frequent regeneration of GAC. Thus, due to the high cost of GAC regeneration, ozonation may lead to significant savings.

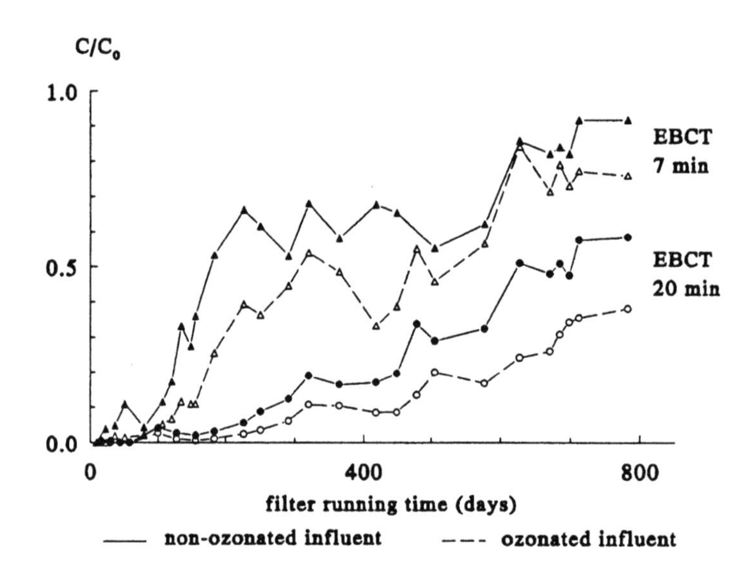

Figure 2.11 Breakthrough of atrazine in GAC filters that received ozonated and non-ozonated pretreated Rhine River water (atrazine was spiked after complete depletion of ozone).

Savings. The following is an estimate of the savings offered by ozonation of GAC filter influent. For this calculation, the design of GAC filters is assumed to be the same with and without ozonation. With such an assumption, ozonation does not change the investment costs of GAC filtration, and brings savings due to less frequent regeneration of GAC only. The actual savings offered by ozonation of GAC filter influent can be calculated precisely if the target compound that determines the frequency of regeneration is known, and if its removal by GAC filtration is established as a function of the ozone dose and EBCT applied. However, lacking such data, a rough estimate is based on the effect that an ozone dose of 0.8 mg/l had on atrazine removal (C_0 = 2.3 μg/l) at EBCT of 20 minutes (Fig. 2.11).

Table 2.3 gives the calculated annual savings that result from less frequent regeneration of GAC receiving ozonated rather than non-ozonated influent. The calculation was done applying various percentages of atrazine breakthrough as the criterion for the regeneration. The GAC bed required is calculated for the annual production capacity of 13×10^6 m^3 (Q_{avg} = 1500 m^3/h), which is planned for the extension of the treatment capacity at the AWS' Leiduin plant, and the empty-bed-contact-time of 20 minutes. The costs of one GAC regeneration, including

GAC make up, are estimated as DFL 650 (US$ 350) per m^3 of GAC. This is about 40% of the cost of virgin GAC applied at Leiduin (Prinsen Geerligs, 1995).

Table 2.3 Annual savings in costs of GAC regeneration offered by ozonation when various atrazine breakthrough percentages are used as the regeneration criteria.

Atrazine breakthrough (%)	Filter running time prior to GAC regeneration (months) without O_3	with O_3	Annual costs of GAC regeneration (DFL) without O_3	with O_3	Annual savings due to ozonation (DFL)
10	9	13	433,000	300,000	133,000
20	13	18	300,000	217,000	83,000
30	17	23	229,000	170,000	59,000

As shown by this table, when 10%-30% breakthrough of atrazine is used as the criterion –and for the EBCT and ozone dose applied– annual regeneration costs are DFL 59,000 - DFL 133,000 lower when GAC filters are receiving ozonated rather than non-ozonated influent. Higher beneficial effect of ozonation is expected at a full scale plant than estimated here. This is because in our experiment, the concentration of atrazine was the same in both influents, while in the full-scale plant, atrazine concentration in ozonated influent will be lower than in non-ozonated influent (see Fig. 2.9). Such reduced atrazine concentration in ozonated influent, caused by oxidation of atrazine, delays the breakthrough of atrazine from GAC filters.

An estimate of the costs involved in the application of ozone is given in Chapter 1. However, in the water treatment concept of AWS, ozonation is applied not only to improve the performance of GAC filters, but also to provide a disinfection barrier. Thus, the savings brought by the improved performance of GAC filters may be considered a clear gain, and need not be reduced for the cost of ozonation.

CONCLUSIONS

Low dose ozonation can be expected to improve removal of pesticides and other organic micropollutants by GAC filtration. This is not only via the well-known effect of ozone-induced oxidation of pesticides, but also due to ozone-induced oxidation of Background Organic Matter (BOM) present in filter influent. Namely, some BOM compounds will be partially oxidized because of ozonation. This will increase their biodegradability, and lower their adsorbability and molecular mass. The increased concentration of biodegradable organic matter results in increased biological activity. This may enhance the biodegradation of organic compounds which are not (partly) oxidized by ozone. In GAC filters receiving ozonated influent also the bioregeneration may play a more pronounced role. Thus, enhanced biodegradation and reduced adsorption of BOM are expected in filters receiving ozonated rather than non-ozonated influent. Both biodegradation and adsorption of pesticides can be improved because of this.

Ozonation of pretreated Rhine River water results in (limited) oxidation of atrazine: about 25%, 45% and 65% of atrazine is oxidized when applying an ozone dose of 0.5 mg/l, 1 mg/l and 1.5 mg/l, respectively. As frequently demonstrated, the resulting lower atrazine concentration in the influent of GAC filters delays the breakthrough of atrazine.

Ozonation of atrazine-containing pretreated Rhine River water results in the formation of desethylatrazine and desisopropylatrazine at concentrations up to 20% and 5% of the original atrazine concentration, respectively. Most likely, other oxidation by-products (*e.g.* hydroxyatrazine, desisopropylatrazine amide, etc.) are also formed. The two by-products measured are more biodegradable but less adsorbable than atrazine. These characteristics make it difficult to predict whether or not the by-products of atrazine oxidation are removed by GAC filtration better than atrazine.

Removal of atrazine by GAC filtration is significantly improved by ozone-induced oxidation of BOM from pretreated Rhine River water. This can be concluded because improved atrazine removal was observed in the filter that received ozonated rather than non-ozonated influent, while atrazine concentration in both influents was the same. Based on the available analytical capacity, removal of atrazine was monitored for two empty-bed-contact-times (7 and 20 minutes). Due to this effect of ozonation, GAC in filters receiving ozonated influent can be regenerated less frequently. This results in important savings.

RESEARCH SCOPE

To identify which of the anticipated processes and relationships given in Figure 2.6 underlie such improved atrazine removal, more detailed investigations were done within this study. Their more specific objectives were the following ones.

To assess the range of ozone doses that may be applied when the role of ozonation is to promote biodegradation in GAC filters, achieve a certain degree of disinfection, and avoid an unacceptable formation of bromate. Such an assessment also defines the range of ozone doses for which experiments need to be conducted. The results are presented in Chapter 3.

To determine the effect of the biodegradation of BOM in GAC filters on the removal of atrazine. In particular, to find out whether improved BOM biodegradation in filters receiving ozonated influent improves biodegradation of atrazine, and whether it improves adsorption of atrazine. The results are presented in Chapter 4.

To quantify the extent to which ozonation reduces the competitive adsorption of BOM. The results are presented in Chapter 5.

To quantify the extent to which ozonation reduces the preloading of BOM. The results are presented in Chapter 6.

To investigate the extent to which the breakthrough of atrazine in GAC filters receiving ozonated influent can be predicted by commonly applied models, such as the simple Adams-Bohart model and the more comprehensive Plug Flow Homogenous Surface Diffusion model. The results are presented in Chapter 7.

ACKNOWLEDGMENTS

Martin Biesterbos, Paul Bonné, Dick van der Gugten, Peter Huis, Joop Janssen, Cees Rotgans, Ed van Soest and Ed Spijkerman constructed the ozonation - GAC filtration pilot plant and additional bench-scale columns used in this research project, maintained them during the operation and contributed to the monitoring of their performance. In addition, and together with many other members of the Amsterdam Water Supply's staff, they provided the most enjoyable working environment at the AWS' Leiduin treatment plant.

SYMBOLS

a_s = external surface area of GAC particle per unit mass of activated carbon ($m^2 \cdot kg^{-1}$)
B = concentration of bacteria that utilize a certain compound ($CFU \cdot m^{-3}$)
b_{max} = maximum utilization rate of a compound per unit of bacteria ($kg \cdot CFU^{-1} \cdot s^{-1}$)
C = concentration of a compound in bulk solution ($kg \cdot m^{-3}$)
C_0 = concentration of a compound in filter influent ($kg \cdot m^{-3}$)
C_s = concentration of a compound in a vicinity of GAC particle ($kg \cdot m^{-3}$)
d = diameter of GAC pores (m)
D_s = surface diffusion coefficient ($m^2 \cdot s^{-1}$)
K = Freundlich coefficient $(kg \cdot m^{-3})^{-n}$
k_f = film mass transfer coefficient ($m \cdot s^{-1}$)
K_s = half-saturation constant for a compound ($kg \cdot m^{-3}$)
M_r = mass of a compound removed in GAC filter (kg)

n = Freundlich coefficient (-)
r = radius of GAC particle (m)
t = time (s)
Q = water flow ($m^3 \cdot s^{-1}$)
q = mass of a compound adsorbed per unit mass of activated carbon (-)
Y = growth yield of bacteria for a certain compound ($CFU \cdot kg^{-1}$)

REFERENCES

Adams, C.D. and S.J. Randtke (1992). Ozonation by-products of atrazine in synthetic and natural waters. *Env.Sci.Tech.*, 26:2218-2227.

AWS (1993). Determination of some nitrogen- and phosphorous-containing pesticides in water. *Water Quality Monitoring Department: Analysis-instruction F-19/L* (in Dutch).

AWS (1995). Internal report.

AWWA Committee (1981). Assessment of microbial activity on GAC. *Journal AWWA*, 8:447-454.

Billen, G., P. Servais, P. Bouillot and C. Ventresque (1992). Functioning of biological filters used in drinking water treatment - the Chabrol model. *J. Water SRT - Aqua*, 4:231-241.

Boere, J. (1992). Advanced treatment processes: activated carbon treatment. IHE, Delft.

Boere, J.A. and R. de Jonge (1996). Personal communication.

Brock, T.D. and M.T. Modigan (1988). *Biology of microorganisms.* Prentice Hall International Inc.

Degrémont (1994). Removal of micropollutants: the case of pesticides. *Proc. Technical symposium on water treatment,* Amsterdam.

DeWaters, J.E. and F.A. DiGiano (1990). The influence of ozonated natural organic matter on the biodegradation of a micropollutant in a GAC bed. *Journal AWWA*, 8:69-75.

Foster, D.M., A.J. Rachwal and S.L. White (1992). Advanced treatment for the removal of atrazine and other pesticides. *Water Supply,* 10:133-146.

Graveland, A. (1994). Application of biological activated carbon filtration at Amsterdam Water Supply. *Water Supply,* 14:233-241.

Graveland, A. and J.P. van der Hoek (1995). Introduction of biological activated carbon filtration at Leiduin. *H_2O,* 19:573-579 (in Dutch).

Graveland, A. (1995). Biological Activated Carbon. In *Proc. INKA Workshop,* IHE, Delft.

Huck, P.M., P.M. Fedorak and W.B. Anderson (1991). Formation and removal of assimilable organic carbon during biological treatment. *Journal AWWA,* 12:69-80.

Kearney, P.C., M.T. Muldoon, C.J. Somich, J.M. Ruth and D.J. Voaden (1988). Biodegradation of ozonated atrazine as a wastewater disposal system. *J.Agric.Food Chem.*, 36:1301-1306.

Knappe, D.R.U. (1996). *Predicting the removal of atrazine by powdered and granular activated carbon.* Ph.D. thesis, University of Illinois, Urbana, USA.

Langlais, B., D.A. Reckhow and D.R. Brink (1991). *Ozone in water treatment,* Lewis Publishers and AWWA Res.Found., Denver, p. 63-79 and 177-207.

Laskin, A.I. and H.A. Lechevalier (1974). *Handbook of microbiology.* CRC Press Inc., Cleveland.

Legube, B., S. Guyon and M. Doré (1987). Ozonation of aqueous solutions of nitrogen heterocyclic compounds. *Ozone Sci.Eng.,* 9:233-246.

Najm, I.N., V.L. Snoeyink and Y. Richard (1991). Effect of initial concentration of a SOC in natural water on its adsorption by activated carbon. *Journal AWWA,* 83:57-63.

Namkung, E. and B.E. Rittmann (1987). Removal of taste- and odor-causing compounds by biofilms grown on humic substances. *Journal AWWA,* 7:107-112.

NORIT (1993). Single solute isotherm protocol.

McCarty, P.L., M. Reinhard and B.E. Rittmann (1981). Trace organics in groundwater. *Env.Sci.Techn.,* 1:41-51.

Olmstead, K.P. and W.J.Jr. Weber (1991). Interactions between microorganisms and activated carbon in water and waste treatment operations. *Chem.Eng.Comm.,* 108:113-125.

Orlandini, E. (1992). *Relating microbial activity to granular activated carbon characteristics.* M.Sc. thesis, IHE, Delft.

Perrotti, A.E. and C.A. Rodman (1974). Factors involved with biological regeneration of activated carbon. *Water,* 144:316-325.

Prinsen Geerligs, W.L. (1995). Design and building of ozonation and GAC filtration at Amsterdam Water Supply. *H_2O*, 19:580-588 (in Dutch).

Smeenk, J.G.M.M., O.I. Snoek and R.C. Lindhout (1990). From Rhine to pure (Van Rijn naar Rein). *H_2O*, 5:126-135 (in Dutch).

Sontheimer, H., J.C. Crittenden and R.S. Summers (1988). *Activated carbon adsorption for water treatment*, AWWA - DVGW Forschungsstelle Engler Bunte Institut, Karlsruhe.

Snoeyink, V.L. (1990). Adsorption of organic compounds. In: *Water quality and treatment.* 4th edition, AWWA, McGraw-Hill, p. 781-875.

Xiaojin, Z., W. Zhansheng and G. Xiasheng (1991). Simple combination of biodegradation and carbon adsorption - the mechanism of the biological activated carbon process. *Water Res.*, 2:165-172.

Yao, C.C.D. and W.R. Haag (1991). Rate constants for direct reactions of ozone with several drinking water contaminants. *Water Res.*, 7:761-773.

Ying, W. and W.J.Jr. Weber (1979). Bio-physicochemical adsorption model systems for wastewater treatment. *J. Water Pollution Control Fed.*, 11:2661-2677.

Van der Kooij, D. and A. Visser (1976).*Removal of organic compounds in a filter filled with activated and in a filter filled with non-activated carbon and the presence and behavior of bacteria in these filters.* Kiwa report SW-164 (in Dutch).

Van der Kooij, D. (1983). Biological processes in carbon filters. *In: Activated carbon in drinking water technology.* Kiwa/AWWA cooperative research report. AWWA Research foundation.

Van der Kooij, D., W.A.M. Hijnen and J.C. Kruithof (1989). The effects of ozonation, biological filtration and distribution on the concentration of easily Assimilable Organic Carbon (AOC) in drinking water. *Ozone Sci. Eng.*, 11:297-311.

Weber, W.J.Jr. (1972). *Physicochemical processes for water quality control.* Wiley & Sons, New York.

Weber, W.J.Jr., M. Pirbazari and G. Melson (1978). Biological growth on activated carbon: an investigation by scanning electron microscopy. *Env.Sci.Tech.*, 7:817-819.

VEWIN (1993). *Recommendations.* VEWIN, Rijswijk, The Netherlands.

Willemse, R.J.N. and J.C. van Dijk (1994). Modeling assimilable organic carbon degradation in biologically active filters. *H_2O*, 2:34-40 (in Dutch).

Chapter 3

Ozonation *vs.* Disinfection and Formation of Biodegradable Organic Matter and Bromate [1]

ABSTRACT— Several pilot plant experiments were done to assess the ozone doses that may be applied for the combination of ozonation and Granular Activated Carbon (GAC) filtration. The objective of these experiments was to relate (i) formation of bromate, (ii) formation of Assimilable Organic Carbon (AOC), and (iii) the Ct (mg O_3/l · min) value achieved to the ozone dose applied.

All the experiments conducted show that ozone doses lower then 0.8 mg/l result in less than 10 µg/l, which is the standard proposed for the European Union. Doses up to 3 mg/l can be applied when taking into account that about 96% of bromate formed will be removed by reverse osmosis, which is planned as the final step for the extension of the treatment capacity at the Amsterdam Water Supply's Leiduin plant.

Ozonation was found to increase the concentration of AOC. The higher the ozone dose, the lower the AOC/O_3 ratio obtained: an ozone dose of 0.75 mg/l increases the concentration of AOC for more than 50% of the maximum 170 µg Ac-C/l that were formed for doses up to 3 mg/l. No pronounced seasonal effect was observed regarding the AOC formed for a given ozone dose.

An ozone dose of 1.5 mg/l may be expected to result in the Ct value sufficient for, at least, 2-log inactivation of viruses and 1-log inactivation of *Giardia* cysts, while a dose of 3 mg/l may be expected to result in at least 14-log and 7-log inactivation of viruses and *Giardia* cysts, respectively. Disinfection in the full-scale plant can be higher, if the short-circuiting in ozone contact tanks can be limited. No pronounced seasonal effect was observed regarding the disinfection credit achieved for a given ozone dose.

---

[1] Published by E. Orlandini, J.C. Kruithof, J.P. van der Hoek, M.A. Siebel and J.C. Schippers (1997) in *J. Water SRT - Aqua*, 1:20-30.

3.1 INTRODUCTION

The multi-functional role of ozonation in the treatment philosophy of Amsterdam Water Supply has been introduced in Chapter 1. One role of ozonation is to provide an essential contribution to the disinfection of water. Another role is to oxidize partly organic matter present in water, and to significantly improve removal of this matter, notably pesticides and other organic micropollutants, by GAC filtration. However, ozonation should not result in an unacceptable formation of bromate. This aspect deserves special attention because, due to high bromide (150-200 µg/l) and low DOC (≈ 2 mg/l) concentration in Rhine River water, the formation of bromate may be expected to limit the ozone doses that may be applied.

To assess the range of applicable ozone doses, several pilot plant experiments were conducted. Their objective was to relate the following three aspects of ozonation to the ozone dose applied to pretreated Rhine River water:

- formation of bromate;
- formation of biodegradable organic matter, measured as the concentration of Assimilable Organic Carbon (AOC);
- disinfection credit achieved, measured as the value of Ct (mg O_3/l·min).

In addition, the extent to which the formation of bromate at the AWS' Leiduin plant can be described by the "molecular ozone" model for bromate formation (Haag and Hoigné, 1983) was also investigated.

3.2 THEORETICAL BACKGROUND

3.2.1 Formation of bromate

Ozonation of bromide-containing water may result in the formation of bromate, a compound considered possibly carcinogenic to humans (Kurokawa *et al.*, 1990). Recently formulated WHO guideline of 25 µg/l, the standard of 10 µg/l proposed for the EU, and the more stringent bromate-standard considered in the Netherlands (see Ch. 1), raised interest in factors controlling the formation of bromate. Already in 1983, Haag and Hoigné proposed a model for bromate formation during ozonation (Fig. 3.1, full lines). Without ammonia, and assuming the negligible formation of brominated organic compounds, this model is reduced to the reactions shown in Fig. 3.1 by bold lines (the "molecular ozone" model). The available rate constants and pK_a incorporated in the "molecular ozone" model have been determined for 20°C and, except for k_1 and pK_a, their temperature dependence is not known (Haag and Hoigné, 1983; Von Gunten and Hoigné, 1992).

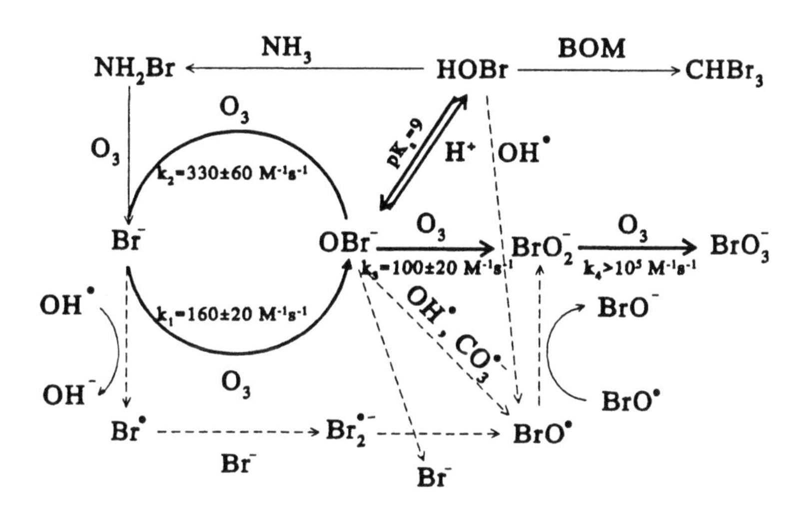

Figure 3.1 Bromate formation model (Von Gunten and Hoigné, 1992).

Assuming a batch or uniform plug flow reactor, the change in concentration of various bromine species over time may be described by the following set of equations:

$$\frac{d[HOBr]_{tot}}{dt} = k_1[O_3][Br^-] - \alpha(k_2+k_3)[O_3][HOBr]_{tot} \quad (3.1)$$

$$\frac{d[Br^-]}{dt} = -k_1[O_3][Br^-] + \alpha k_2[O_3][HOBr]_{tot} \quad (3.2)$$

$$\frac{d[BrO_3^-]}{dt} = \alpha k_3[HOBr]_{tot}[O_3] \quad (3.3)$$

$$K_a = \frac{[H^+][OBr^-]}{[HOBr]} \quad (3.4)$$

$$[HOBr]_{tot} = [OBr^-] + [HOBr] \quad (3.5)$$

$$\alpha = \frac{[OBr^-]}{[HOBr]_{tot}} \quad (3.6)$$

The reaction of BrO_2^- and O_3 to BrO_3^- (Fig. 3.1) has a much higher rate constant (>10^5 $M^{-1}s^{-1}$) than other reactions of the "molecular ozone" model. Thus, it is considered to occur instantaneously and has not been represented in the system of equations given above.

The importance of radical reactions (Fig. 3.1, dashed lines) for the formation of bromate increases in cases of enhanced ozone decomposition into OH radicals ($OH^{\cdot}$). This normally occurs at an elevated pH or when H_2O_2 is added, but the formation of OH radicals may also be accelerated by high water temperature and specific organic compounds present in water (Hoigné, 1988).

Von Gunten and Hoigné (1994) and Von Gunten *et al.* (1996) developed the model for the formation of bromate that includes not only the reactions of molecular ozone, but also the reactions of OH and CO_3 radicals. This model was found to accurately predict the formation of bromate under real treatment conditions (Von Gunten *et al.*, 1995). However, the concentration of OH radicals had to be determined from the measured oxidation of atrazine, for which the rate of reactions with both ozone and OH radicals is known (Hoigné, 1997). A tracer compound (atrazine) was needed because the concentrations of OH radicals are too small to be measured directly and, also, cannot be predicted. Many attempts have been made so far to predict the formation of OH radicals, but it is not likely that a unique concept for this purpose can be found.

Decomposition of ozone into OH radicals was found to lower the bromate formed for the same ozone dose applied (Von Gunten and Hoigné, 1994; Von Gunten *et al.*, 1996; Meijers and Kruithof, 1993). Therefore, adding H_2O_2 to promote decomposition of ozone to OH radicals is a possible treatment option when aiming to reduce the formation of bromate, especially if the objective of ozonation is oxidation of pesticides such as atrazine. Note that H_2O_2 needs to be added at sufficiently high H_2O_2/O_3 ratios in order to avoid possible synergistic effect causing an increased bromate formation (Von Gunten and Hoigné, 1994; Kruithof *et al.*, 1995). However, H_2O_2 addition is not beneficial when the objective of ozonation is disinfection. Decomposition of ozone to OH radicals, which have poor disinfection properties but contribute to bromate formation, results in more bromate formed for a given extent of disinfection achieved (Von Gunten *et al.*, 1996). When aiming to disinfect water while keeping the formation of bromate as low as possible, a much better treatment strategy is to lower the pH of water. This slows the decomposition of ozone. Thus, it increases the disinfection achieved for a given ozone dose. At the same time, lower pH reduces the formation of bromate.

3.2.2 Formation of biodegradable organic matter

All of the methods applied for determining the concentration of biodegradable organic matter in treated drinking water are bioassays. The reasons are difficulties one encounters when trying to identify a variety of organic compounds present in water and quantify them at low concentrations. As reviewed by Huck in 1990, there are two groups of these methods. One group measures the growth of bacteria in water samples incubated with indigenous or pure bacterial cultures. Another group measures the reduction in DOC concentration upon this incubation. The factor that limits the application of DOC-based methods is the low precision of currently applied analyzers (≈100 μg C/l). In contrast, the methods based on the measurement of the extent of bacterial growth may have much higher precision (≈1 μg C/l).

Assimilable Organic Carbon (AOC) determination has been developed for the measurement of the bacterial regrowth potential. It is based on the measurement of the maximum extent of growth (maximum colony count, N_{max}) of a selected pure bacterial culture, in representative water samples in which the indigenous bacteria have been killed or inactivated by heat treatment (Van der Kooij *et al.*, 1982; Van der Kooij, 1990). The concentration of AOC is calculated using the N_{max} value and the yield coefficient of the organism for a selected substrate. In most of the investigations so far, and in this study, the concentration of AOC was determined from the growth of *Pseudomonas fluorescens* strain P17 and *Spirillum species* strain NOX. Normally, the two cultures are inoculated as a mixture in samples of water to be tested, and the AOC concentration is calculated from their maximum colony counts and growth yield on acetate.

The above two organisms prefer different groups of compounds. P17 has great nutritional versatility, and may grow on a variety of amino acids, carbohydrates and aromatic acids. It can also grow on carboxylic acids, but not on formic, glyoxylic and oxalic acids that are typical by-products of ozonation. In contrast to P17, strain NOX can use a wide range of carboxylic acids –including the three mentioned above– but it cannot use carbohydrates, alcohols, or aromatic acids. Strain NOX can also use a few amino acids; however, this organism does not assimilate amino acids when growing on mixtures of compounds. Therefore, the growth of strain NOX may be used for the determination of the concentration of carboxylic acids in water (Van der Kooij, 1990).

As reviewed by Langlais *et al.* (1991) and Huck *et al.* (1991), many researchers have shown that ozonation increases the concentration of biodegradable organic matter in water. The same conclusion was drawn irrespective of the method applied to determine the concentration of biodegradable organic matter. It appears that no additional increase in the concentration of biodegradable organic matter occurs above a certain ozone dose: beyond 1-2 mg O_3/mg DOC all biodegradable organic carbon likely to be formed has already been transformed (Langlais *et al.*, 1991). In this context, it may also be noted that Huck *et al.* (1991) observed an attenuation in the growth of P17 in ozonated water. They hypothesized that either ozonation produces compounds that are inhibitory for the growth of P17, or that strain NOX competes more successfully for the substrate available.

3.2.3 Disinfection

The Surface Water Treatment Rule (SWTR) applied in the USA defines the Ct values required as a function of the extent of disinfection, the organism to be inactivated, disinfectant applied and water temperature (Von Huben, 1991). The Ct value is the product of the time of exposure (min) and the residual disinfectant concentration (mg/l) measured at the end of this time. Furthermore, the SWTR defines the time of exposure used for Ct calculation as the time at which 10% of water introduced at the inlet of the tank has passed to its outlet (t_{10}). Obviously, for plug flow (perfectly baffled tanks) t_{10} equals the hydraulic detention time in the tank (t), while for ideally mixed flow (unbaffled tanks) t_{10} equals 10% of the detention time. Utilities are required to conduct tracer studies to determine the exact t_{10} valid for their conditions. Only in exceptional cases, a rule-of-thumb may be used to relate the t_{10}/t ratio (0.1-1) to the baffling condition in the tank.

Because of the definition of Ct value, a higher Ct value can be claimed when ozone concentration is measured as a function of time rather than only at the end of the total contact time. To increase the Ct value that may be claimed, the SWTR allows utilities to measure residual disinfectant concentration during the total contact time as frequently as technically feasible. Therefore, the actual Ct values that may be claimed in a full-scale installation are determined not only by ozone dose and the rate of its depletion, but also by the design of the ozonation step itself. The design will determine the total time available for ozone contact, the intervals of time at which residual ozone concentration can be measured (normally at the outlet of each serial contact tank), and the short-circuiting in ozone contact tanks.

The Ct values required for various degrees of inactivation of viruses and *Giardia* cysts by ozone are given in Table 3.1 (Von Huben, 1991). There is presently insufficient information available to determine the efficiency with which ozonation inactivates oocysts of *Cryptosporidium*, which is another disease-causing protozoan commonly found in surface water. However, the Ct values required appear to be higher than those required for inactivation of *Giardia* cysts. As shown by this table, a lower Ct value is required at higher water temperature. The reason is an increase in the efficiency of disinfection with temperature, since higher water temperatures speed the rate of the diffusion of ozone through cell walls and the reaction of ozone with key enzymes. However, this does not mean that a lower ozone dose is required at higher water temperatures. Higher water temperatures also speed up the rate of both ozone induced oxidation of organic matter and ozone decomposition to OH radicals and oxygen and, consequently, they reduce the ozone exposure achieved for a given ozone dose.

Table 3.1 Ct values (mg O_3/l·min) for inactivation of *Giardia* cysts and viruses (Von Huben, 1991).

Temperature (°C)	≤ 1	5°C	10°C	15°C	20°C	25°C
Giardia inactivation						
1-log	0.97	0.63	0.48	0.32	0.24	0.16
2-log	1.9	1.3	0.95	0.63	0.48	0.32
3-log	2.9	1.9	1.43	0.95	0.72	0.48
Viruses inactivation						
2-log	0.9	0.6	0.5	0.3	0.25	0.15
3-log	1.4	0.9	0.8	0.5	0.4	0.25
4-log	1.8	1.2	1.0	0.6	0.5	0.3

3.3 MATERIALS AND METHODS

3.3.1 Experimental setup

Table 3.2 Water quality in various experiments.

Date	T (°C)	pH	DOC (mg/l)	[HCO_3^-] (mg/l)	UV_{254} (1/m)	turbidity (FTU)	[NH_3] (µg N/l)	[Br^-] (µg/l)
10/93	12	8.0	2.0	153	6.1	0.25	40	162
01/94	8	8.0	2.1	162	6.4	0.13	30	--
04/94	10	8.0	1.9	154	5.4	0.18	30	160
06/94	18	7.9	2.5	169	5.1	0.18	--	172
08/94	20	7.9	2.1	158	5.9	0.15	<20	182
10/94	14	7.9	2.0	172	6.2	0.20	<20	258
02/95	9	7.7	2.1	147	5.4	0.16	<20	115
08/95	24	7.5	1.6	149	4.4	0.13	<20	176

By coagulation, sedimentation and rapid sand filtration pretreated Rhine River water was ozonated in a counter-current ozonation column (height 5 m,diameter 0.1 m) of a pilot plant operated at the AWS' Leiduin plant. Ozone was produced from pure oxygen, at a gas-concentration of about 30 mg O_3/l. Ozone doses up to 4 mg/l were applied, which was achieved by varying the flow of both water (250-1000 l/h) and the ozone gas (10-40 l/h). The transfer efficiency of the ozone dosing column was determined as at least 95%, and the transferred ozone dose was assumed to equal that applied. To account for changes in the temperature and composition of water, several experiments were conducted over a period of two years (Table 3.2). The concentrations of bromate and AOC in the effluent of the ozonation column were determined after complete depletion of ozone, without the addition of a reducing agent.

3.3.2 Analytical methods

The analytical methods applied are standardized methods used by Amsterdam Water Supply. Ammonium and DOC concentrations were measured in a continuous flow system (AA-2, Technicon) by colorimetric analysis (720 and 550 nm, respectively), with a detection limit of 0.05 mg NH_4^+/l and 0.1 mg C/l. HCO_3^- concentration was determined by automatic titration with HCl. The liquid and gas phase ozone concentrations were measured by the DPD (diethyl-P-phenylenediamine) method and iodometric method, respectively (APHA, 1980). Unless stated otherwise, Assimilable Organic Carbon concentrations were measured in duplicate, applying the simultaneous incubation of strains P17 and NOX (Van der Kooij, 1990).

Bromate was measured applying ion chromatography (AG9-SC guard and AS9-SC analytical column, Dionex, Sunnyvale, CA, USA) with 2000 µl injection loop (Smeenk *et al.*, 1994). Eluent containing 0.7 mM $NaHCO_3$ was used for ion separation, followed by a gradient step to purge the column (5 mM $NaHCO_3$/5 mM Na_2CO_3). Bromate was detected by both UV detection (200 nm) and chemically suppressed conductivity measurement. In chemically suppressed conductivity detection, a device is placed between the analytical column and the detector. Its purpose is to inhibit detector response to the ionic constituents in the eluent other than the ions of interest, so as to lower the detector background, and at the same time enhance the response of the detector to the ions of interest (*e.g.* bromate). For this method, a membrane suppressor Dionex ASRS was applied. The detection limit of this method was 0.05 µg BrO_3^-/l for demineralized water. The method was found free from matrix interferences for the samples investigated. A calibration (five standards) was done before and after each series of samples, and control standards were verified every five injections. Bromide was analyzed by ion chromatography, applying the same guard and analytical column as for bromate analysis, but using different carbonate eluent (1.4 mM Na_2CO_3/0.2 mM $NaHCO_3$). Bromide was detected by both UV detection (200 nm) and chemically suppressed conductivity measurement, with a detection limit of 5 µg/l.

3.3.3 Bromate formation model and Ct calculation

About twenty 50 ml samples of the effluent of the ozonation column were taken for each ozone dose tested. They were used to measure the residual ozone concentration during the period required for the complete depletion of ozone (ozone residual < 0.01 mg/l). The residual concentrations of ozone at any given point in time were required for the modeling of bromate formation and for the calculation of Ct value. They were determined from the following exponential function that was fitted to the residual ozone concentrations measured (Richard, 1994):

$$O_3(t) = O_3(0) \cdot e^{-nt} \quad (3.7)$$

For the prediction of bromate formation only –thus not for the calculation of Ct values– residual ozone concentrations in the ozonation column were also taken into account. They were assumed to be constant during the whole hydraulic detention time in the ozonation column, and to equal 50% of the ozone concentrations measured in the column effluent. This assumption is justified for counter-current ozonation columns (USEPA, 1989). Depending on the experiment and ozone dose, the detention time in the ozone dosing column was between 2 minutes and 12 minutes. Although no tracer studies were done, the column geometry allowed the assumption of plug flow conditions.

The equations of the "molecular ozone" model (Eq. 3.1 - Eq. 3.6) were solved numerically, via a finite differences-based program written for that purpose. The average values shown by Fig. 3.1 were applied for the rate constants (k_1, k_2 and k_3), while the pK_a of 9 was assumed for hypobromous acid (Von Gunten and Hoigné, 1994).

3.4 RESULTS AND DISCUSSION

3.4.1 Formation of bromate

Hundreds of micrograms of bromate were found formed by ozone doses up to 4 mg/l (Fig. 3.2). In contrast to the results of Kruithof *et al.* (1995), no pronounced seasonal effect was observed. Bromate formed by a particular ozone dose is given in Table 3.3.

Table 3.3 Effect of ozonation on bromate concentration in treated water.

ozone dose (mg O_3/l)	0	0.75	1.5	3.0
[1]BrO_3^- (µg/l)	0	<10	<150	<250
[1,2]BrO_3^- (µg/l)	0	< 0.4	< 6	< 10

[1] formed upon the complete depletion of ozone
[2] 96% removal of bromate by reverse osmosis taken into account

Ozone doses lower then 0.8 mg/l were found to result, in all experiments, in less bromate than the proposed European guideline of 10 µg/l. When the removal of about 96% of bromate by reverse osmosis is taken into account (Van der Hoek *et al.*, 1995b), 250 µg/l of bromate may be produced by ozonation without exceeding the concentration of 10 µg/l in the finished water. Therefore, as shown by Table 3.3, ozone doses up to 3 mg/l may be applied.

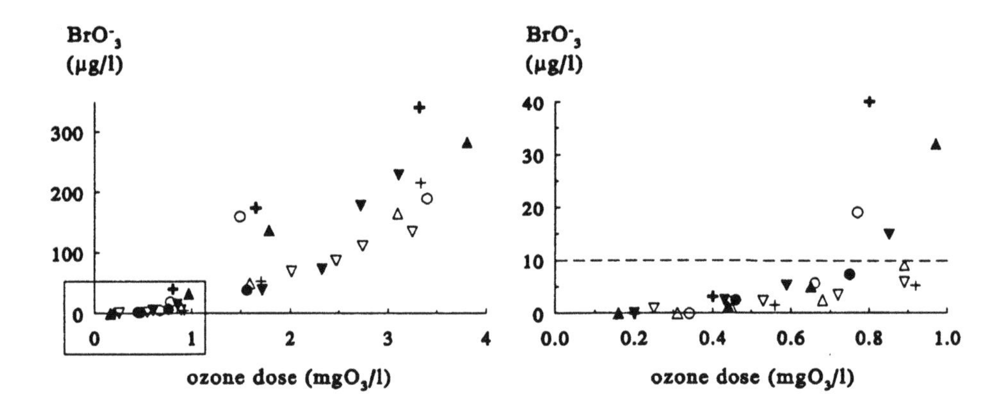

Figure 3.2 Bromate formation as a function of the ozone dose. The whole graph (left) and the enlarged initial part of the graph (right).

The higher ozone doses may be applied, if required, because full-scale ozonation may be expected to result in less bromate than shown by Fig. 3.2. First, the formation of bromate can be reduced by decreasing the pH of water, since the experiments were conducted at pH values as high as 8.0. Note that low pH does not conflict with the objectives of ozonation, which are disinfection and promotion of biological activity in GAC filters, rather than pesticide (*e.g.* atrazine) oxidation. Secondly, bromate forming reactions in these experiments were allowed to take place until all residual ozone was depleted and they took place under ideal plug-flow conditions (in the counter-current ozonation column and sample bottles) which maximizes the formation of bromate. On the other hand, in the proposed extension of the AWS' Leiduin plant, GAC filtration will quench residual ozone after about 30 minutes retention time and, depending on the short-circuiting in the ozone contact tanks, the conditions will deviate from the ideal plug flow.

For more accurate determination of the quantity of bromate that may be expected in practice, the experiments should be conducted under the same conditions as those applied in the full-scale installation, or the formation of bromate should be modeled. Successful modeling of bromate formation should allow accurate prediction for any considered retention time.

3.4.2 Modeling bromate formation

Given the concentration of ozone as a function of time (Fig. 3.3) and the initial Br^- concentration (Table 3.2), the formation of bromate was modeled by the "molecular ozone" model (Haag and Hoigné, 1983). To quantify the formation of bromate via the reactions of OH radicals some experiments were conducted with and without the addition of 0.1mM of t-butanol which scavenges these radicals. This t-butanol dose was found sufficient, since the addition of higher concentrations of t-butanol did not further decrease the bromate formed by ozone doses up to 4 mg/l.

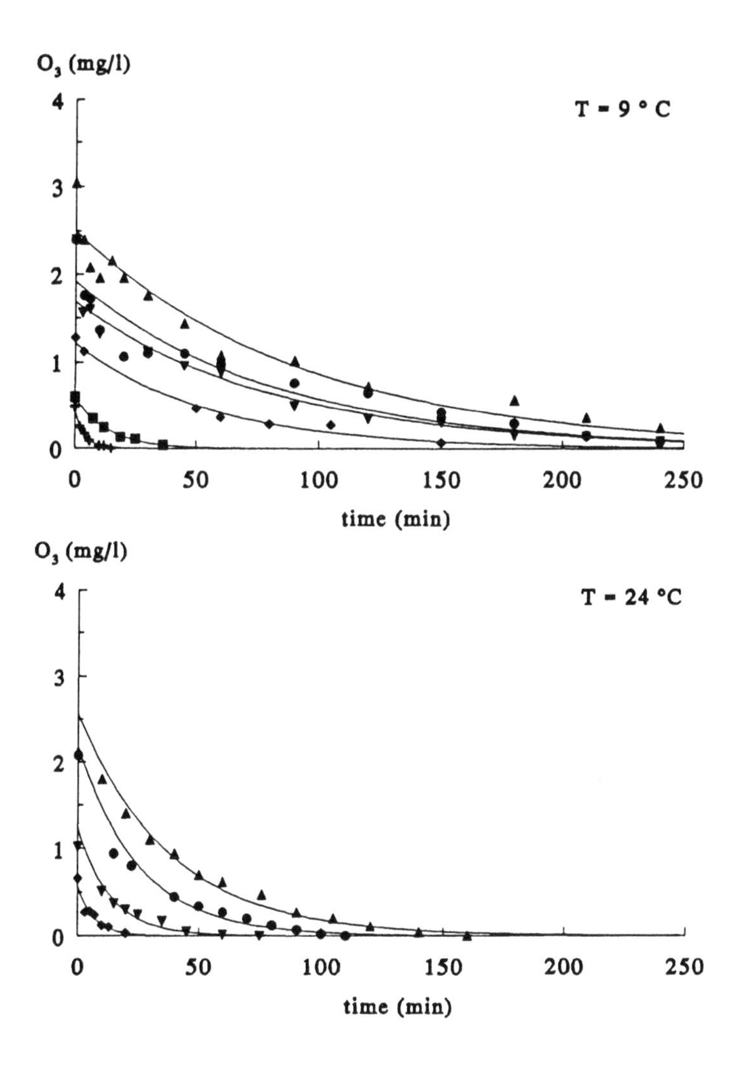

Figure 3.3 Ozone depletion at low and high water temperature. Measured ozone residuals (points) and exponential functions fitted to these data (lines).

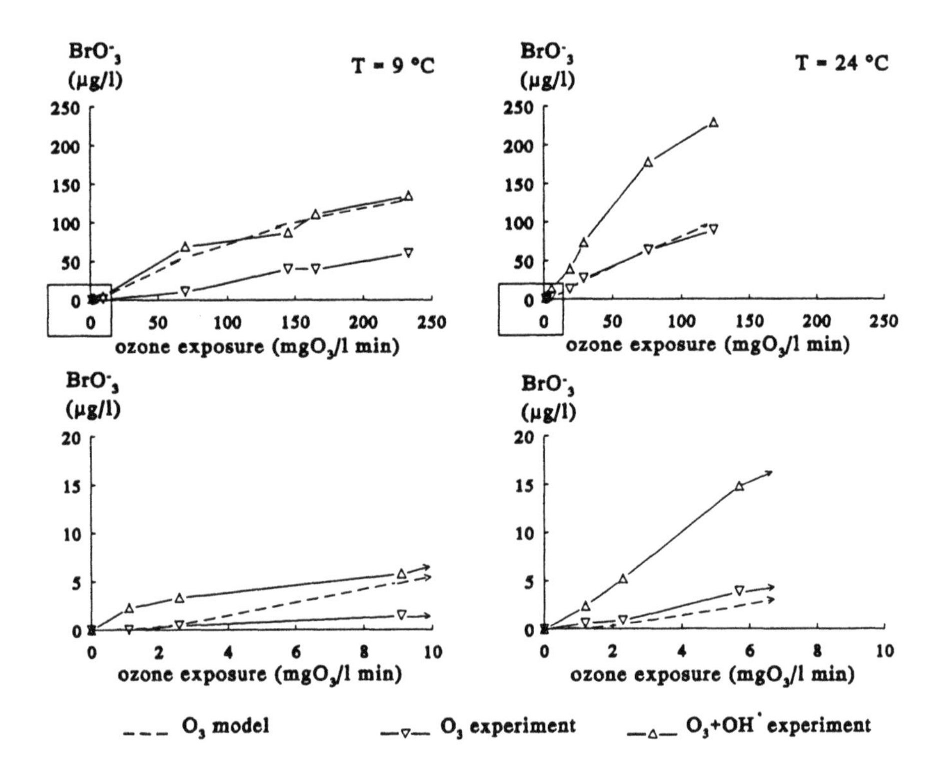

Figure 3.4 Bromate formation at low and high water temperature: prediction of molecular ozone model (O_3 model), bromate formed via molecular ozone reactions only (O_3 experiment), and bromate formed via both O_3 and $OH^•$ reactions (O_3+$OH^•$ experiment).

Fig. 3.4 gives the results of the experiments conducted in February and August 1995. A similar trend was observed in previous experiments as well (Orlandini *et al.*, 1994; Van der Hoek *et al.*, 1995a). The results clearly show that more bromate is being formed when no t-butanol is added, which means that the formation of bromate proceeds via both O_3 and $OH^•$ mediated reactions. Therefore, the "molecular ozone" model –which does not account for the formation of bromate via OH radical reactions– does not allow accurate predictions for the conditions at AWS. This is consistent with the results obtained at other water works (Von Gunten and Hoigné, 1994; Von Gunten *et al.*, 1996). This is a serious drawback, because the formation of bromate by OH radicals cannot be predicted unless their concentration is estimated from the measured oxidation of some probe compound (Von Gunten *et al.*, 1995; Hoigné, 1997).

However, Fig. 3.4 also shows that at water temperature of approximately 20°C, the "molecular ozone" model accurately predicts the bromate formed via O_3-reactions only. In contrast, this model predicts higher bromate formation via O_3-reactions than measured at 10°C. Such a discrepancy may be expected because the model operates with the rate constants and the pK_a of hypobromous acid, which have been determined at 20°C (Haag and Hoigné, 1983; Von Gunten and Hoigné, 1992). These constants are expected to have lower values at 10°C than at 20°C, which should reduce the bromate formed for a given ozone exposure.

3.4.3 Formation of biodegradable organic matter

Ozonation was found to increase the concentration of biodegradable organic matter in water, measured as the Assimilable Organic Carbon (Fig. 3.5). No pronounced seasonal effect was observed. Raising the ozone dose, the ratio between the AOC formed and the ozone dose applied went down: 0.75 mg O_3/l (O_3/DOC ≈ 0.3) resulted already in about 50% (95 µg Ac-C/l) of the maximum AOC that can be formed (170 µg Ac-C/l), while no clear increase in AOC concentration was observed when the ozone dose was raised from 1.5 to 3 mg/l.

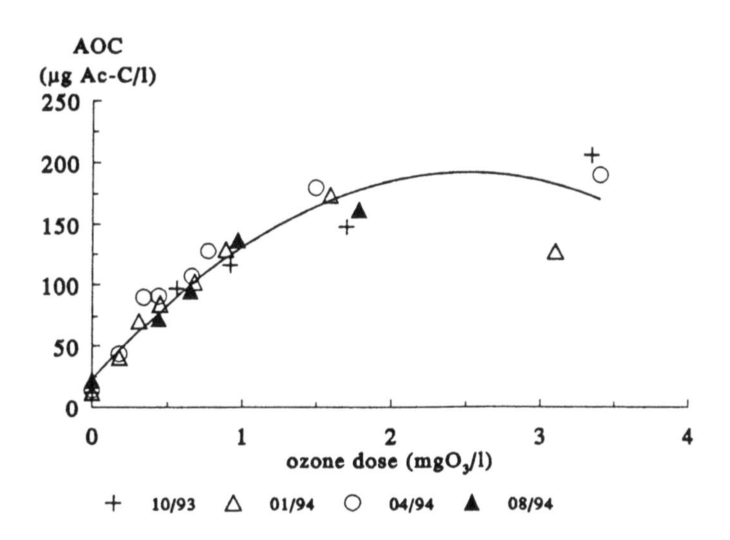

Figure 3.5 AOC formation as a function of the ozone dose.

In June 1994, the concentration of AOC in water ozonated with 0.5 mg O_3/l was determined (i) by applying the simultaneous incubation of both strains P17 and NOX, (ii) by applying first the incubation of P17 followed by the incubation of NOX when P17 had reached maximum colony count, and (iii) by applying first the incubation of NOX followed by the incubation of

P17 when NOX had reached the maximum colony count (Fig. 3.6). The results of this single experiment are in good agreement with the results of Van der Kooij *et al.* (1989). They suggest that carboxylic acids account for more than 80% of the AOC in ozonated water (determined as the ratio between AOC_{NOX} and AOC_{total} obtained when NOX was incubated before P17) and that about 40% of these are oxalic, glyoxylic and formic acid (determined as the ratio between AOC_{NOX} obtained when NOX was incubated after and before P17). Note that the concentrations of AOC in Fig. 3.6 were determined from the growth yield of NOX on acetate only, while this strain has about four times lower growth yield on oxalate than on acetate: $2.9 \cdot 10^6$ and $12 \cdot 10^6$ CFU/µg-C, respectively (Van der Kooij, 1990). Therefore, if oxalic acid –which is very resistant to molecular ozone (Hoigné and Bader, 1983)– was assumed to be the main carbon source for NOX, these percentages would be even higher.

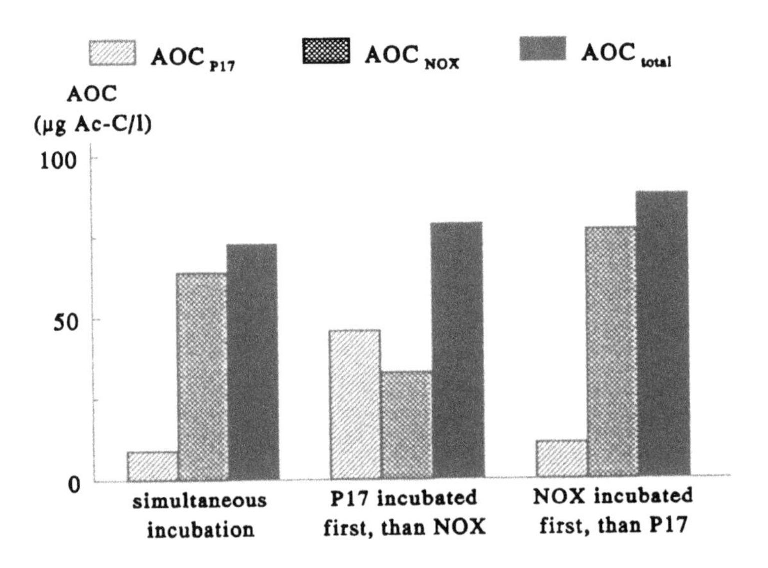

Figure 3.6 Growth of P17 and NOX in ozonated water when they are incubated simultaneously of one after another.

Four times more colonies of P17 were observed when it was incubated first, than when it was incubated after or simultaneously with NOX (Fig. 3.6). This suggests that the strain NOX compete better for the substrate available in ozonated water than P17. Thus, the attenuation in the growth of P17, which was observed at AWS when applying ozone doses up to 1 mg/l (Orlandini *et al.*, 1994), is most likely caused by more successful competition of NOX for the carboxylic acids present in ozonated water, and not by the presence of inhibitory compounds formed by ozonation.

3.4.4 Disinfection

Given the hydraulic detention time in ozone-contact tanks and the residual ozone concentration as a function of time, the maximum Ct value (Ct_{max}) may be claimed when ozone concentrations in a full-scale installation may be measured at minute intervals of time, and when the ozone contact tanks are perfectly baffled (uniform plug flow, $t_{10}/t = 1$). The Ct value equals then the integral of the ozone depletion function over the hydraulic detention time. On the other hand, the minimum Ct value (Ct_{min}) may be claimed when the residual ozone concentration can be measured only at the end of the total contact time, and when the ozone contact tanks are not baffled at all (ideally mixed flow, $t_{10}/t = 0.1$). The Ct value equals then 10% of the product of the hydraulic detention time in ozone-contact tanks and the residual ozone concentration measured at the end of that time.

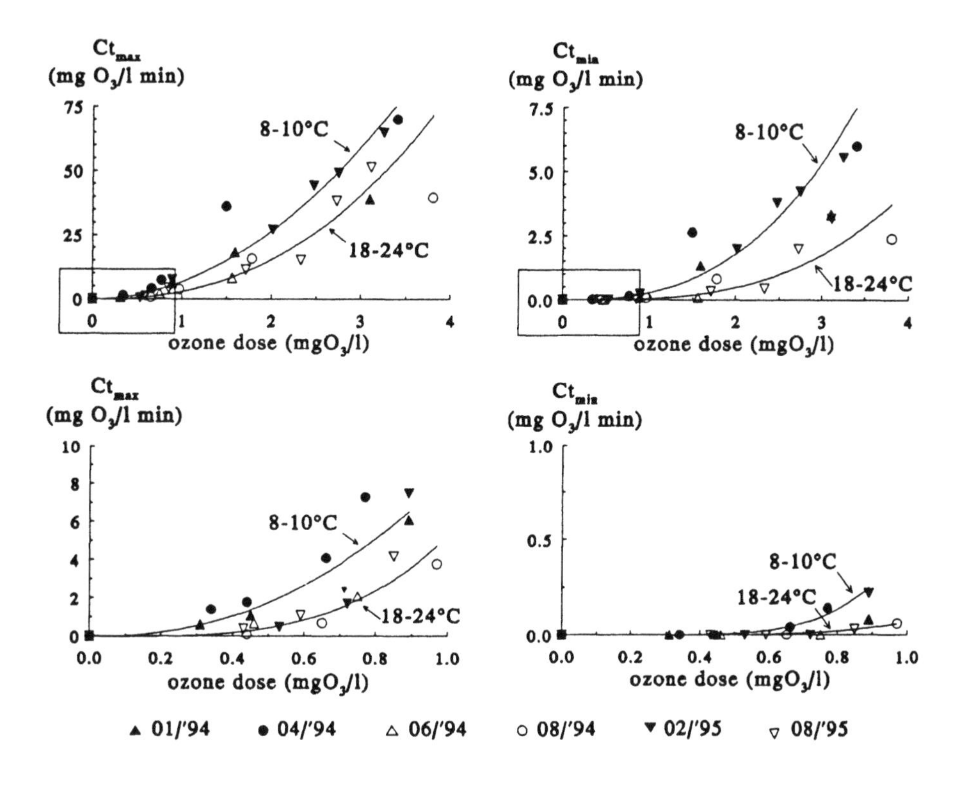

Figure 3.7 Ct_{max} (left graphs) and Ct_{min} (right graphs) values determined for 30 minutes contact time, as a function of the ozone dose and temperature.

In the proposed extension of the Leiduin water treatment plant, the anticipated hydraulic detention time between ozonation and GAC filtration –which quenches ozone– is about 30 minutes. Therefore, the Ct_{max} values were calculated integrating the ozone depletion functions over the initial 30 minutes (Fig. 3.7, left), while Ct_{min} values were calculated as 10% of the product of 30 minutes and the residual ozone concentration measured after this time (Fig. 3.7, right). Although not shown here, ozone depletion functions similar to those shown by Fig. 3.3 were determined for all experiments, and were used for this calculation.

Table 3.4 gives the Ct_{min} and Ct_{max} values that may be expected for a particular ozone dose, and the log-inactivations of viruses and *Giardia* cysts that may be expected for these Ct values. Further investigation is needed to determine the actual Ct value that may be claimed in the full-scale installation. This investigation should take into account the extent of baffling in ozone-contact tanks, and the intervals at which ozone residuals can be measured during the total ozone-contact time.

Table 3. 4 Effect of ozonation on disinfection credit achieved.

ozone dose (mg O_3/l)	0.75	1.5	3.0
[1,2]Ct_{min} - [1,3]Ct_{max} @ 10°C	0.1 - 3.9	0.8 - 15	5.4 - 60
[1,2]Ct_{min} - [1,3]Ct_{max} @ 20°C	0.0 - 1.5	0.3 - 8	1.8 - 41
[4]log reduction viruses @ 10°C	0.5 - 15	3 - 60	21 - 240
[4]log reduction viruses @ 20°C	0.1 - 12	2 - 64	14 - 328
[4]log reduction *Giardia* @ 10°C	0.2 - 8	1.7 - 31	11 - 125
[4]log reduction *Giardia* @ 20°C	0.0 - 6	1 - 33	7 - 170

[1] calculated for 30 minutes contact time (mg O_3/l·min)

[2] assumed ozone measurement only at the end of contact time and the ideally mixed flow

[3] assumed ozone measurement at minute intervals of contact time and the uniform plug flow

[4] which may be claimed for the above determined Ct_{min} and Ct_{max} values

The higher water temperature resulted in a pronounced reduction of the Ct value achieved for a given ozone dose. The reason for this is an increase in the rate of ozone depletion at higher temperatures (Fig. 3.3), that may be explained as due to faster oxidation of organic matter by ozone, and faster decomposition of ozone to oxygen and OH radicals. However, this reduction of the Ct value had only a limited effect on the disinfection credit achieved for a given ozone dose, because the same log inactivation requires lower Ct values at higher water temperatures.

CONCLUSIONS

As demonstrated in all experiments, ozone doses lower then 0.8 mg/l result in less than 10 µg/l of bromate, which is the standard proposed for the European Union. Taking into account the removal of about 96% of bromate by reverse osmosis, which is planned as the final step in the extension treatment scheme, ozone doses up to 3 mg/l will result in less than 10 µg/l of bromate in the finished water.

The formation of bromate can be reduced by decreasing the pH, given that the experiments were conducted at pH values as high as 8, and that low pH does not conflict the objectives of ozonation (being disinfection and partial oxidation of the organic matter present in water). Furthermore, it is expected that less bromate will be formed in the full-scale plant than estimated above. The reason is that bromate forming reactions were allowed to act until the complete depletion of ozone, and that they took place under ideal plug flow conditions, neither of which will be the case in the full-scale plant.

The "molecular ozone" model proposed by Haag and Hoigné in 1983 does not allow accurate predictions for the conditions at AWS, because bromate formation by OH radicals turned out to be essential. This is consistent with the results obtained at other water works (Von Gunten and Hoigné, 1994; Von Gunten *et al.*, 1996). The "molecular ozone" model accurately predicted bromate formed at 20°C via O_3-reactions only, but it predicted more bromate than actually formed via O_3 reactions at 10°C. The most likely reasons for this discrepancy are the constants with which this model operates: they have been determined at 20°C, and the model does not account for their temperature dependency.

Ozonation increases the concentration of biodegradable organic matter in water, measured as the Assimilable Organic Carbon (AOC). But, the higher the ozone dose, the lower the AOC/O_3 ratio obtained: a dose of 0.75 mg/l ($O_3/DOC \approx 0.3$) increases the concentration of AOC for about 95 µg Ac-C/l –more than 50% of the maximum 170 µg Ac-C/l that could be formed– while there is no clear difference between the concentrations of AOC produced by ozone doses of 1.5 mg/l and 3 mg/l. No pronounced seasonal effect is observed regarding the effect of ozonation on AOC formation.

The results suggest that carboxylic acids account for more than 80% of the AOC in ozonated water. A low dose ozonation (< 1 mg/l) does not appear to produce compounds that are inhibitory for the growth of strain P17: the observed attenuation in the growth of P17 in ozonated water may be attributed to the more successful competition of strain NOX for the substrate available.

When complete mixing in ozone contact tanks is assumed, an ozone dose of 1.5 mg/l may be expected to result in a Ct value that is sufficient for, at least, 2-log inactivation of viruses and 1-log inactivation of *Giardia* cysts. A dose of 3 mg/l may be expected to result in at least 14 and 7-log inactivation of viruses and *Giardia* cysts, respectively. Further investigation is needed to determine precisely the extent of disinfection that may be achieved by full-scale ozonation: the inactivation of viruses and *Giardia* cysts can be higher when measures are taken to limit the short-circuiting (mixed flow) in ozone contact tanks.

Higher water temperatures accelerate ozone depletion and, as a result, reduce the Ct value achieved for a given ozone dose. However, because the same inactivation of viruses and *Giardia* requires lower Ct value at higher temperature, water temperature only has a limited effect on the disinfection credit achieved for a given ozone dose.

ACKNOWLEDGMENT

I am very much indebted to Ed Spijkerman (AWS), who helped –most skillfully– to conduct the experiments described in this chapter. I also thank Willem van Rossum and Erwin van den Ende (both AWS) who measured bromate and AOC concentrations, respectively.

REFERENCES

APHA (1980). *Standard methods for the examination of water and wastewater.* 15th ed., APHA, Washington DC.

Haag, W.R. and J. Hoigné (1983). Ozonation of bromide-containing waters: kinetics of formation of hypobromous acid and bromate. *Env.Sci.Tech.*, 17:261-267.

Hoigné, J. and H. Bader (1983). Rate constants of reactions of ozone with organic and inorganic compounds in water. *Water Res.*, 17:185-194.

Hoigné, J. (1988). The chemistry of ozone in water. In: Stucki S., ed. *Process technologies for water treatment.* Plenum Publishing Corporation.

Hoigné, J. (1997). Inter-calibration of OH radical sources and water quality parameters. *Wat. Sci. Techn.*, 35:1-8.

Huck, P.M. (1990). Measurement of biodegradable organic matter and bacterial growth potential in drinking water. *Journal AWWA*, 7:78-86.

Huck, P.M., P.M. Fedorak and W.B. Anderson (1991). Formation and removal of assimilable organic carbon during biological treatment. *Journal AWWA*, 12:69-80.

Kruithof, J.C., E.J. Oderwald-Muller and R.T. Meijers (1995). Control strategies for the restriction of bromate formation. *Proceedings IOA World Congress*, Lille, Vol. 1, p. 209-221.

Kurokawa, Y., A. Maerkawa, M. Taqkahashi and Y. Hayashi (1990). Toxicity and carcinogenicity of potassium bromate-a new renal carcinogen. *Env. Health Pers.*, 87:309-335.

Langlais, B., D.A. Reckhow and D.R. Brink (1991). *Ozone in water treatment.* Denver, Lewis Publishers and AWWA Research Foundation.

Meijers, R.T. and J.C. Kruithof (1993). Potential treatment options for restriction of bromate formation and bromate removal. *Proceedings IWSA International Workshop on bromate and water treatment*, Paris, p. 199-206.

Orlandini, E., J.C. Kruithof, M.A. Siebel and J.C. Schippers (1994). Assessing ozonation as pretreatment for GAC filtration. *Proceedings IOA Regional Conference*, Zurich, p. 333-345.

Richard, Y. (1994). Ozone water demand test. *Ozone Sci.Eng.*, 16:355-365.

Smeenk, J.G.M.M., W.J.M. van Rossum, W.J.H. Gademan and C. Bruggink (1994). Trace level determination of bromate and chlorite in drinking and surface water by ion chromatography with automated preconcentration. *Proceedings International Ion Chromatography Symposium*, Torino.

USEPA (1989). *Guidance manual for compliance with the filtration and disinfection requirements for public water systems using surface water sources.* USEPA, Washington DC.

Van der Hoek, J.P., E. Orlandini, A. Graveland and J.G.M.M. Smeenk (1995a). Minimizing bromate formation during ozone-activated carbon treatment. *Proceedings IWSA World Congress*, Durban, pp. SS5-1 - SS5-7.

Van der Hoek, J.P., P.A.C. Bonné, W.J.M. van Rossum and E. Spijkerman (1995b). Removal of bromate by reverse osmosis. H_2O, 1:23-26 (in Dutch).

Van der Kooij, D., A. Visser and W.A.M. Hijnen (1982). Determining the concentration of easily assimilable organic carbon in drinking water. *Journal AWWA*, 10:540-545.

Van der Kooij, D., W.A.M. Hijnen and J.C. Kruithof (1989). The effects of ozonation, biological filtration and distribution on the concentration of easily assimilable organic carbon (AOC) in drinking water. *Ozone Sci.Eng.*, 11:297-311.

Van der Kooij, D. (1990). Assimilable organic carbon (AOC) in drinking water. In: McFeters GA, ed. *Drinking water microbiology: progress and recent developments.* Springer-Verlag.

Von Gunten, U. and J. Hoigné (1992). Factors controlling the formation of bromate during ozonation of bromide-containing waters. *J Water SRT - Aqua,* 41:2911-304.

Von Gunten, U. and J. Hoigné (1994). Bromate formation during ozonation of bromide containing waters: interaction of ozone and hydroxyl radical reactions. *Env.Sci.Techn.*, 28:1234-1242.

Von Gunten, U., J. Hoigné and A. Bruchet (1995). Oxidation in ozonation processes: application of reaction kinetics in water treatment. *Proceedings IOA World Congress*, Lille, Vol. 1, pp. 17-25.

Von Gunten, U., A. Bruchet and E. Costentin (1996). Bromate formation in advanced oxidation processes. *Journal AWWA,* 88:6:53-65.

Von Huben, H. (1991). *Surface water treatment: the new rules.* AWWA, Denver.

Chapter 4

Biodegradation and Atrazine Removal by Granular Activated Carbon Filtration

ABSTRACT —More efficient atrazine removal was observed in the Granular Activated Carbon (GAC) filter that received ozonated rather than non-ozonated pretreated Rhine River water. Various pilot and bench scale experiments were done to verify that such improved removal is due to better biodegradation and/or better adsorption of atrazine, and that both are increased by enhanced biodegradation of Background Organic Matter (BOM) after ozonation. No indication of atrazine biodegradation in the GAC filters operated was found in either of the four different experiments conducted. Therefore, biodegradation of atrazine in GAC filters and its better biodegradation in filters receiving ozonated influent were not confirmed. However, the results obtained do not entirely exclude the possibility that atrazine was biodegraded in the GAC filters operated. If atrazine biodegradation in these filters was indeed negligible, it is possible to assume that the conditions for the growth of atrazine degrading bacteria were inadequate.

The enhanced biodegradation of BOM in filters receiving ozonated influent improves adsorption of atrazine in GAC filters. This can be concluded because atrazine was found better adsorbed onto GAC preloaded with ozonated water that passed through Non-Activated Carbon (NAC) filters, than onto GAC preloaded directly with ozonated water. Because of negligible adsorption of BOM in NAC filters, only biodegradation of BOM in NAC filters could account for the improved adsorption of atrazine that was observed. However, reduced adsorbability and molecular mass of ozonated BOM compounds can also contribute to the improved adsorption of atrazine in GAC filters.

4.1 INTRODUCTION

Granular Activated Carbon (GAC) filtration preceded by a low dose ozonation, the joint process termed Biological Activated Carbon filtration (Miller and Rice, 1978), was found to remove atrazine better than GAC filtration alone (see Ch. 2). The influent was Rhine River water pretreated with coagulation, sedimentation and rapid sand filtration.

It is well-known that ozone-induced oxidation of atrazine decreases its concentration in the influent of GAC filters and, by that, delays its breakthrough from these filters (Degrémont, 1994; Foster *et al.*, 1992). However, in our experiment, atrazine concentration was the same in both influents because atrazine was spiked after all dosed ozone was depleted. Consequently, improved atrazine removal in filters receiving ozonated influent can be the result of improved biodegradation and/or improved adsorption of atrazine. Both may be expected because of enhanced biodegradation of Background Organic Matter (BOM) in these filters. More BOM biodegradation enlarges the density of bacteria, which may also degrade atrazine. In addition, increased BOM removal via biodegradation rather than adsorption can reduce both the competitive adsorption and the preloading of BOM, resulting in better adsorption of atrazine.

The aim of the work presented in this chapter is to explain the role of biodegradation of BOM and atrazine in the removal of atrazine in GAC filters. More specifically, the objectives are to verify that:

- ozonation increases biodegradation of BOM in GAC filters;
- ozonation increases biodegradation of atrazine in GAC filters;
- biodegradation of ozonated BOM reduces its preloading onto GAC, resulting in improved adsorption of atrazine.

The effect of BOM biodegradation on its competitive adsorption was not investigated in this study.

The major difficulty encountered when trying to verify biodegradation of either BOM or atrazine in GAC filters is how to distinguish clearly between their biodegradation and adsorption in these filters (AWWA, 1981). As a result, the three aforementioned objectives could not be achieved directly, by simply comparing the performance of GAC filters receiving ozonated and non-ozonated influent, but the following assumptions had to be made.

For the first objective, we assumed that increased biodegradation of BOM in GAC filters receiving ozonated influent can be verified by showing improved removal of BOM in Non-Activated Carbon (NAC) filters receiving ozonated rather than non-ozonated influent. The same DOC concentration is expected in both non-ozonated and ozonated influent, because ozone doses of up to 4 mg O_3/l were not found to reduce DOC concentration in water (data not presented here). Thus, because of negligible BOM adsorption in NAC filters, improved BOM removal in NAC filters receiving ozonated influent can be attributed to enhanced BOM biodegradation in these filters. If it occurs in NAC filters, higher biodegradation of ozonated than of non-ozonated BOM may be expected to occur in GAC filters also. Such expectation is based on the assumption that activated carbon does not provide worse support for microbial growth than non-activated carbon.

For the second objective, we assumed that biodegradation of atrazine in GAC filters, and its improved biodegradation in filters receiving ozonated influent, can be verified by showing one or more of the following effects:

a) The presence of one or more of the metabolites of atrazine in the effluent of atrazine-spiked GAC filters (after an empty-bed-contact-time of either 7 minutes or 20 minutes) and the higher concentration of one or more of these metabolites in the effluent of the filters receiving ozonated water. Note that atrazine needs to be spiked after ozonation, to avoid ozone-induced formation of by-products of atrazine oxidation.

b) Better atrazine removal in GAC filters when biodegradable organic matter is added to their influent. Addition of biodegradable organic matter is not expected to improve adsorption of atrazine. Thus, better atrazine removal (if observed) can be attributed to increased biodegradation of atrazine.

c) The removal of atrazine in Non-Activated Carbon (NAC) filters, and its improved removal in NAC filters receiving ozonated influent. Because of negligible atrazine adsorption in NAC filters, atrazine removal can only be attributed to its biodegradation. If it occurs in NAC filters, atrazine biodegradation may be assumed to occur in GAC filters also.

d) The removal of atrazine in liquid media inoculated with bacteria taken from atrazine-spiked GAC filters. Such a result shows the presence of "atrazine degraders" in these filters and, therefore, indicates biodegradation of atrazine in them. The same result in the batch and column experiment with the known "atrazine degraders", which were also conducted, indicates that biodegradation of atrazine can be shown under the experimental conditions applied, and that at least a certain type of atrazine degrading bacteria can grow in filters.

e) More atrazine removed from water and less atrazine adsorbed onto GAC in the pilot plant filter receiving ozonated rather than non-ozonated atrazine-spiked influent.

The third objective is to verify whether biodegradation of BOM in GAC filters receiving ozonated influent reduces its preloading, resulting in better adsorption of atrazine. We assumed that this can be done by showing that atrazine is better adsorbed onto GAC preloaded with ozonated water that has first passed through filters filled with non-activated carbon, than onto GAC preloaded directly with ozonated water. If such better atrazine adsorption is observed, it can only be caused by BOM biodegradation in NAC filters, which may be expected to occur in GAC filters also.

4.2 THEORETICAL BACKGROUND

4.2.1 Biodegradation of BOM in GAC filters and bioregeneration of GAC

As discussed in Chapter 2, besides adsorption, biodegradation of organic compounds also takes place in GAC filters. In the ongoing discussion on the mechanism of biodegradation in GAC filters (AWWA, 1981; Van der Kooij, 1983; Olmstead and Weber, 1991; Orlandini, 1992; Billen *et al.*. 1992; Graveland and Van der Hoek, 1995; Graveland, 1995) it is generally agreed that bacteria and other microorganisms that colonize GAC filters may use organic matter that is dissolved in water. The various models that have been developed (Ying and Weber, 1979; Billen *et al.*, 1992; Willemse and Van Dijk, 1994) propose that bacteria utilize compounds at a rate described by the Michaelis-Menten kinetics. Therefore, the extent of biodegradation in GAC filter is increased by higher concentration of biodegradable compounds in filter influent, their higher biodegradability and higher water temperature. At the same time, the extent of biodegradation in GAC filter is also increased by longer contact time in the filter and use of GAC with high affinity for attachment and low affinity for detachment of bacteria.

Biodegradation of adsorbed compounds, termed bioregeneration of GAC, is highly beneficial because it restores adsorption capacity of GAC. All the mechanisms by which bacteria can utilize compounds adsorbed onto GAC are presently not well defined (see Ch. 2). However, the mechanisms proposing biodegradation of adsorbed compounds upon their desorption from the pores of GAC has nowadays been widely accepted (AWWA, 1981; Sontheimer *et al.*, 1988; Snoeyink, 1990). Desorption of adsorbed compounds may occur due to several causes. First, it may occur due to an increased water temperature that shifts adsorption equilibrium. Secondly, desorption may occur due to reduced concentration of these compounds in water passing through the GAC filter. Such reduced concentration may be caused by biodegradation of these compounds in GAC filters, or by normal (*eg.* seasonal) fluctuations of water quality. Thirdly, previously adsorbed compounds may be displaced due to competitive adsorption of compounds that are more adsorbable with respect to GAC.

Graveland and Van der Hoek (1995) observed at the Amsterdam Water Supply's Weesperkarspel treatment plant an increased specific (per mass of DOC removed) oxygen consumption in GAC filters at higher water temperatures. Based on this finding they postulated that biodegradation of previously adsorbed compounds, termed bioregeneration of GAC, plays a pronounced role. Similar observations have also been made in the late seventies at Bremen and Mülheim water works (Sontheimer *et al.*, 1988). Note that the actual extent of bioregeneration can be overestimated due to other phenomena which may play a role, such as increased removal of oxygen by physicochemical processes during the initial phase of filter operation or increased endogenous respiration of bacteria at high water temperature (AWWA, 1981; Van der Kooij and Visser, 1976).

As reviewed by Huck *et al.* (1991), numerous researchers have shown that ozonation increases the concentration of biodegradable organic matter in water. The same conclusion was drawn irrespective of the method applied for the determination of the concentration of biodegradable organic matter. Consequently, ozonation can be expected to increase the extent of BOM biodegradation in GAC filters. Such expectation was first confirmed by the results of the pilot- and full-scale testing at the Mülheim plant in Germany where ozonation was shown to improve the removal of organic matter in GAC filters significantly, measured both as UV-absorbance and Dissolved Organic Carbon content (Sontheimer *et al.*, 1978). Since then similar observations have also been made elsewhere (Sontheimer *et al.*, 1988; Graveland, 1994).

Graveland (1994) postulated that the improved removal of organic matter in GAC filters receiving ozonated rather than non-ozonated influent was not only due to the increased concentration of biodegradable organic matter in the GAC filter influent. He asserted that the resulting increased biological activity also enhances biodegradation of organic compounds which are not (partly)oxidized by ozone and that the capacity for biodegradation of organic matter adsorbed on GAC (bioregeneration of GAC) is strongly increased. Therefore, ozone-induced oxidation of BOM may be hypothesized to enhance biodegradation of atrazine and other micropollutants in GAC filters.

4.2.2 Biodegradation of atrazine

Many factors determine whether a compound will be biodegraded in the environment. First, the structure (presence and absence of certain substitutes, steric factors, molecular size and other structural features) and the concentration of the compound. Secondly, the type and number of available microorganisms. Thirdly, environmental factors such as pH, temperature, and presence or absence of oxygen and nutrients.

Biodegradation of pesticides may occur via two enzymatic metabolisms: co-metabolism, also called secondary utilization (Namkung and Rittmann, 1987), and catabolism. In co-metabolism pesticides do not serve as an energy source, while in catabolism they do. Depending on the enzymes required for pesticide degradation, co-metabolism may be subdivided in a metabolism via:

- generally present broad-spectrum enzymes such as hydrolase, reductase and oxidase;
- generally present specific enzymes;
- enzymes used only for substrates with structures similar to those of pesticides.

Atrazine is chemically synthesized and its symmetrical triazine structure was foreign to the biosphere when it was first manufactured (Cook, 1987). Such a novel structure lowers the probability that microorganisms are equipped with enzymes required for its degradation. This may explain why such microorganisms have not been readily found throughout the environment, but only on sites where atrazine has been used for many years (Erickson and Kyung, 1989). High persistence of atrazine in surface water, where its degradation occurs via photolysis and via biodegradation, was shown by Schottler and Eisenreich (1994) for the Great Lakes and by Ulrich *et al.* (1994) for various Swiss lakes. While for atrazine half-life-times in surface water are longer than 100 days at 20°C, for readily biodegradable chlorophenoxy herbicides such as 2,4-D (2,4-dichlorophenoxyacetic acid) or MCPA (4-chloro-2-methylphenoxyacetic acid) they are only one to two weeks (WHO, 1993). Shorter half-life-times of at least 20 days to 50 days were reported for atrazine degradation in soil, which proceeds via chemical hydrolysis and via biodegradation (WHO, 1993). Stuyfzand and Lüers (1997) found rapid degradation of atrazine in soil (half-life < 30 days) under anoxic conditions only ($O_2 < 0{,}5$ mg/l; $NO_3 < 0{,}5$ mg/l). Under aerobic conditions ($O_2 \geq 1$ mg/l) they observed rather slow degradation of atrazine (half-life > 120 days).

Beside the novel structure, another possible obstacle for biodegradation of atrazine is its low energy content. The carbon atoms of the triazine ring are at the oxidation level of CO_2, while the side chains are short (Cook, 1987). Nevertheless, atrazine was shown to be used as a sole energy and carbon source, thus via catabolism (Yanze-Kontchou and Gschwind, 1994; Stucki *et al.*, 1995). This was done at high atrazine concentrations of 15 mg/l to 30 mg/l, which are not widely encountered. Atrazine was also shown to be used via co-metabolism, when additional energy and carbon sources such as sucrose, citrate, glycerol and/or mannitol were added. Atrazine was used as a source of either nitrogen (Mandelbaum *et al.*, 1993; Leeson *et al.*, 1993; Assaf and Turco, 1994) or carbon (Behki *et al.*, 1993).

The experimental evidence for the pathway of complete atrazine biodegradation has been summarized by Wackett (1996). Atrazine is biodegraded either via N-dealkylation, with desethylatrazine and desisopropylatrazine as the initial by-products, or via dechlorination with hydroxyatrazine as the initial by-product (Fig. 4.1). While the first two reactions (to desethyl- and desisopropylatrazine) require oxygen, hydroxyatrazine has been reported as the initial by-product of both aerobic and anaerobic degradation of atrazine (Crawford *et al.*, 1997). Biodegradation of atrazine further results in a number of by-products, namely: desisopropylhydroxyatrazine, 2,4-dihydroxy-6-(n'-ethyl)amino-1,3,5-triazine, desisopropyldesethylatrazine, 2-chloro-4-hydroxy-6-amino-1,3,5-triazine, 2-chloro-4,6-dihydroxy-1,3,5-triazine, 2-hydroxy-4,6-diamino-1,3,5-triazine and 2,4-dihydroxy-6-amino-1,3,5-triazine. Irrespective of the initial byproduct, the last in the chain of byproducts for which the triazine ring is still not broken is always cyanuric acid. Cyanuric acid is further degraded to biuret, urea and eventually to HCO_3^-, NH_3 and H^+.

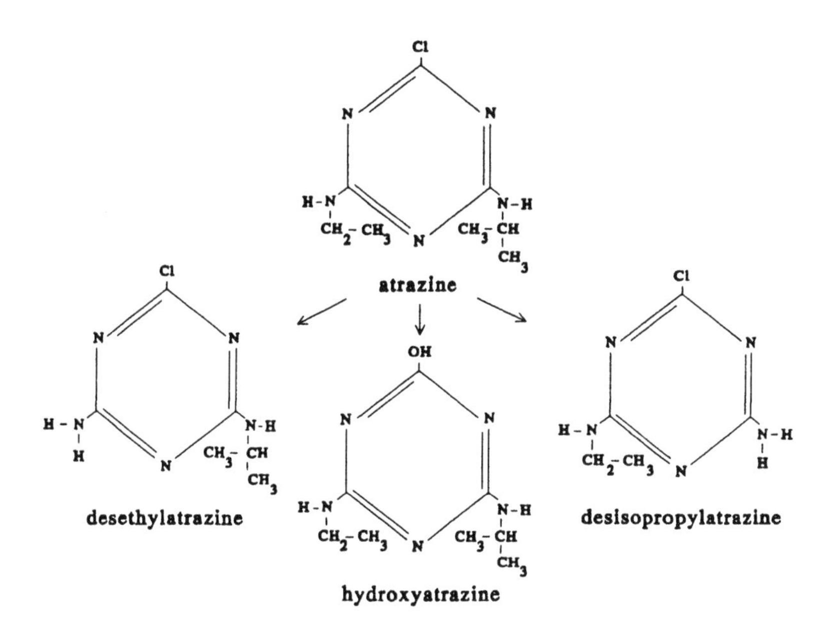

Figure 4.1 The initial by-products of atrazine biodegradation.

4.2.3 Biodegradation of atrazine in GAC filters

Biodegradation of atrazine under drinking water treatment conditions is most likely to proceed via co-metabolism, in which bacteria growing on biodegradable organic matter present in GAC filter influent simultaneously use atrazine as an additional source of carbon or nitrogen (Matsumura, 1982). This is due to low atrazine concentrations of, typically, a few micrograms per liter at most (Schottler *et al.*, 1994; Ulrich *et al.*, 1994).

By ozonation increased concentration of biodegradable organic matter is expected to result in enhanced biodegradation of atrazine via co-metabolism only if the enzymes involved are generally present and used for degradation of at least some biodegradable compounds that result from ozone-induced oxidation of Background Organic Matter (Matsumura, 1982). Whether these conditions will be met is difficult to predict, because the enzymes involved in biodegradation of atrazine have not yet been identified. When biodegradation of atrazine in GAC filters occurs via co-metabolism, the highest rate may be expected in the part of the GAC filter bed where the rate of degradation of easily biodegradable organic matter is the highest (Matsumura, 1982).

Another mechanism for biodegradation of atrazine in GAC filters is catabolism, in which microorganisms use atrazine as the main energy and carbon source. When this mechanism occurs, bacteria are expected to first use compounds that are rich energy sources, and only after their depletion will they start to degrade poor energy sources like atrazine. This is called sequential utilization (Brock and Modigan, 1988). Due to the presence of easily biodegradable organic matter in the influent of GAC filters, sequential utilization of atrazine may require long empty-bed-contact-times (EBCTs) to ensure that easily biodegradable organic matter has been depleted.

The problem posed by sequential utilization is of particular relevance for GAC filters that receive ozonated influent. Namely, biodegradation in GAC filters that are operated with commonly applied EBCTs is not expected to bring the concentration of Assimilable Organic Carbon in the effluent of filters receiving ozonated influent below 10 μg Ac-C/l, which is the value for the final product water recommended by the Netherlands Water Works Association. This makes the concentration of easily biodegradable compounds, even in the effluent of GAC filters, at least 100 times higher than the European Union's standard for a single pesticide (0.1 μg/l), and at least 20 times higher than the standard for all the pesticides combined (0.5 μg/l). It was hypothesized that by low dose ozonation and longer EBCTs in GAC filters (*e.g.* 40 min), bio-activity in GAC filters would be enhanced and maybe atrazine biodegradation would be introduced (Graveland, 1994). However, in all likelihood, much longer EBCTs than this need to be applied. Namely, during the treatment of Meuse River water (AOC concentration 88-199 μg Ac-C/l; temperature 4-21 °C) EBCTs of 20 minutes and 40 minutes were shown to result in effluent AOC concentrations that are above 20 μg/l Ac-C/l and 10 μg Ac-C/l, respectively (Rodman *et al.*, 1995). AOC concentrations lower than those were observed only during the initial stages of filter operation when, presumably, AOC was removed by adsorption also. Similar observation has been made during the experiments at the AWS's Leiduin plant, where EBCTs of 20 minutes and 44 minutes resulted in effluent AOC concentrations above 15 μg Ac-C/l and 10 μg Ac-C/l, respectively (AWS, 1996). The influent of the pilot plant GAC filter used for these experiments was ozonated (0.4-3 mg O_3/l) pretreated Rhine River water with AOC concentrations ranging from 36 µg Ac-C/l to 130 µg Ac-C/l. However, in the full scale plant –where effluents of various GAC filters with different filter running times are combined– lower AOC values than these are expected. During the period 1996-1997, the average AOC concentration after 1st and 2nd step GAC filtration at Leiduin plant was 13 µg Ac-C/l and 7,6 µg Ac-C/l, respectively (AWS, 1996). GAC filtration influent at Leiduin is pretreated and ozonated Rhine River water (AOC ≈ 70 µg Ac-C/l), while each GAC filtration step is operated with an EBCT of 20 minutes.

Bouwer and McCarty (1982) showed that not all micropollutants are biodegraded in GAC filters: while several chlorinated benzenes present in trace concentrations of 10 µg/l to 30 µg/l were readily biodegraded in acetate-spiked (1 mg/l) GAC filters, more recalcitrant chlorinated aliphatics such as chloroform, tetrachloroethylene and 1,1,1-trichloroethane were not. From the results reported so far, concluding firmly whether atrazine will be biodegraded in GAC filters used for drinking water treatment is not possible. Its long half-lives in the environment, where biodegradation is only one degradation mechanism, and its poor removal by slow sand filtration (Foster *et al.*, 1992), do not suggest prompt biodegradation.

On the other hand, Feakin *et al.* (1995) reported biodegradation of atrazine in the GAC filters spiked with 2 mg/l of atrazine. Note that this atrazine concentration is few orders of magnitude higher than atrazine concentrations normally encountered in drinking water sources. Formation of metabolites of atrazine was not reported. Biodegradation of atrazine was quantified as the difference between the mass of atrazine removed in the filters and desorbed from GAC. Feakin *et al.* determined the recovery of their desorption method (97%) after GAC and atrazine were in contact for only 48 hours, but claimed the start of atrazine biodegradation only after filters were operated for two or more weeks. Thus, considering that the recovery of the desorption method decreases in time (Bandjar, 1996), it is possible that reduced recovery of desorption method was mistaken for the biodegradation of atrazine. This can also explain the same extent of atrazine biodegradation reported for the GAC filters inoculated with atrazine degrading bacteria and for the filters that were not inoculated.

Of particular interest are the results of Huang and Banks (1996). Performing their experiments with labeled atrazine, they clearly demonstrated atrazine biodegradation in the GAC columns they operated. The GAC filter influent in these experiments was Kansas River water (TOC = 14 mg/l). About 62% of ring-labeled atrazine was converted to CO_2 in the columns that received ozonated water and ozonated atrazine. It can be derived that up to about 85% of total atrazine-carbon was converted to CO_2 when the carbon in side chains is also accounted for. About 50% of the influent atrazine was converted to CO_2 in the columns that received atrazine and ozonated Kansas River water, while 38% of influent atrazine was converted to CO_2 in columns that received atrazine and non-ozonated Kansas River water. Ozone/air gas mixture was bubbled through water for two hours prior to start of the experiment with GAC columns. In this way GAC filter influent was exposed to 3000 mg O_3/l. Atrazine concentration in the influent of GAC columns was 50 μg/l. The influent was spiked with nutrients and buffer salts (65 mg/l KH_2PO_4, 15 mg/l $CaCl_2 \cdot H_2O$, 215 mg/l K_2HPO_4, 3 mg/l $MgSO_4$ and 250 mg/l $NaHPO_4$). The filter-columns were filled with 2,5 g of ground GAC (average particle diameter 0,5 mm) and glass beads as supporting media. The empty-bed-contact-time (EBCT) in the part of the column filled with GAC was about 42 minutes (assuming $\rho_{GAC} = 0.5$ g/cm^3), while the EBCT in the part filled with glass beads was 7,1 hours. The filtration velocity of 1.4 m/h was achieved by the recycle ratio of 40:1. The columns were sterilized prior to the experiment and were than inoculated with the microbial inoculum taken from the sediment and water of the Turtle Creek Reservoir. This reservoir had a long-time exposure to atrazine and in 1990, when this study was conducted, atrazine concentration in its water was about 50 μg/l.

Finally, it may be noted that atrazine is expected to be biodegraded in GAC filters under aerobic conditions. This is especially the case in GAC filters used for the treatment of drinking water because these filters are allowed to produce anaerobic effluent only during their initial start-up when oxygen in intensely chemisorbed onto GAC. However it is not impossible (although it is not likely) that local anaerobic conditions may occur in these GAC filters during their further operation. For local anaerobic conditions to occur, the rate of oxygen utilization by bacteria needs to be higher than the rate of oxygen diffusion from bulk solution. If this condition is met, atrazine biodegradation (if occurring) may also proceed via the anaerobic pathway.

4.3 MATERIALS AND METHODS

4.3.1 Biodegradation of BOM

An ozonation - GAC filtration pilot plant was operated at the Amsterdam Water Supply's Leiduin treatment plant (Fig. 4.2). The plant comprised four GAC filters, two receiving non-ozonated and two receiving ozonated (0.8 ± 0.2 mg O_3/l) Rhine River water pretreated by coagulation, sedimentation and rapid sand filtration. The filters were filled to a bed depth of 1.1 m with the extruded activated carbon ROW 0.8S manufactured by NORIT NV. Before the filters were put into operation, ultra-pure water (*i.e.* permeate of the reverse osmosis plant) was percolated through them for 11 days (v = 1 m/h), after which they were backwashed to ensure removal of fine materials and stratification of the GAC bed. This washing of GAC for 11 days also eliminated problems associated with the potential increase in pH and the precipitation of calcium carbonate onto GAC particles (AWS, 1995) and, especially, any difference in this respect between the GAC filters receiving ozonated and non-ozonated influent. Concentration of Background Organic Matter, measured as Dissolved Organic Carbon (DOC) concentration, was monitored during the first two years of plant operation. This was done in both ozonated and non-ozonated influent of the GAC filters, and in their effluents at a bed depth of 0.35 and 1.1 m. These two bed depths correspond with the empty-bed-contact-times (EBCTs) of 7 minutes and 20 minutes. These two EBCTs were chosen for monitoring based on considerations regarding the expected breakthrough of atrazine. The EBCT of 7 minutes was chosen because it allows clear breakthrough of atrazine (ca. 50%) in a relatively short period of time (ca. 6 months). This allows prompt testing of the hypothesis that ozonation improves removal of atrazine in GAC filters and, if required, prompt adjustment of the research scope (see. Ch. 2). The EBCT of 20 minutes was chosen because it has been applied at several water treatment plants. Moreover, longer EBCTs (for instance 40 minutes) would require several years of filter operation for atrazine to break through.

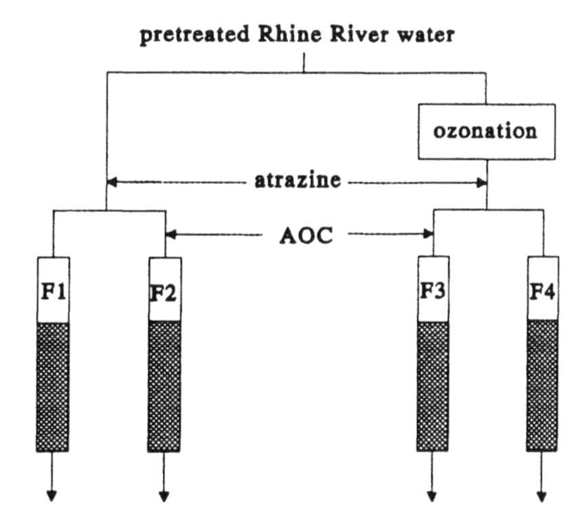

Figure 4.2 Scheme of the ozonation - GAC filtration pilot plant operated at the Amsterdam Water Supply's Leiduin plant.

In addition to GAC filters, two filters filled with non-activated carbon (NAC) were operated for 200 days (from January to August of 1994). One filter received ozonated and one non-ozonated influent (pretreated Rhine River water). The filters had an inner diameter of 5 cm and were operated with an EBCT of 20 minutes. NAC particles were of the same shape and size as the particles of the GAC used (NORIT ROW 0.8S). DOC concentration was measured in the influent and effluent of the two NAC filters (EBCT = 20 min) throughout their operation.

Dissolved Organic Carbon (DOC) concentration was measured in a continuous flow system (AA-2 from Technicon) by colorimetric analysis at 550 nm. The detection limit of this method is 0.1 mg-C/l. Before the measurement of DOC concentration, samples were filtered through glass-fibre-filters with pore openings of 0.7 μm.

4.3.2 Biodegradation of atrazine

Atrazine and two among its by-products, desethylatrazine and desisopropylatrazine, were measured via an analytical method that involves liquid-liquid extraction with ethyl-acetate, and gas chromatography with nitrogen-phosphorous detection (see Ch. 2). Hydroxyatrazine and other atrazine by-products were not measured because the analytical method and the required equipment were not available.

Presence of atrazine metabolites. The influent of the aforementioned pilot plant GAC filters (Fig. 4.2) was spiked with 2.2±0.2 μg/l of atrazine. Ozonated influent was spiked after complete depletion of ozone. During the initial two years of pilot plant operation, atrazine, desethylatrazine, and desisopropylatrazine were regularly monitored in the influents of the filter that received non-ozonated water (F1) and the filter that received ozonated water (F4), and in their effluents at a bed depth of 0.35 (EBCT 7 min) and 1.1 m (EBCT 20 min). The number of samples collected totaled 550.

Addition of biodegradable organic matter. One pilot plant filter receiving non-ozonated influent (F2) and one receiving ozonated influent (F3) were spiked with biodegradable organic matter (Fig. 4.2). Two experiments were conducted. In the first experiment (water temperature 20°C, DOC 1.8 mg/l), 110 μg C/l of acetate was added to the influent of each filter. In the second experiment (water temperature = 25°C, DOC = 2.0 mg/l) 250 μg-C/l of oxalate, 30 μg-C/l of acetate, 10 μg-C/l of glyoxal and 10 μg-C/l of formaldehyde were added. The compounds besides acetate were chosen as the typical byproducts of ozonation of naturally occurring organic matter (Van der Kooij and Hijnen, 1984; Krasner *et al.*, 1993; Weinberg *et al.*, 1993). The 250 μg/l of oxalate-C added corresponds to 60 μg/l of acetate-C equivalents, because the growth yield on oxalate is about four times lower than the growth yield on acetate. Such a difference in the growth yield was shown for the *Spirillum species* strain NOX that uses both compounds (Van der Kooij, 1990). Lacking the data for the other two compounds, we assumed that 10 μg/l of both glyoxal-C and formaldehyde-C equal 10 μg/l of acetate-C equivalents. Thus, the mixture added equaled 110 μg/l of Ac-C equivalents.

Atrazine removal in NAC filters. While the two Non-Activated Carbon (NAC) filters were used to monitor the removal of BOM (see objective 1), they were also spiked with

2.7±0.6 μg/l of atrazine. Ozonated influent was spiked after complete depletion of ozone. Atrazine, desethylatrazine, and desisopropylatrazine concentrations were monitored in the influent and effluent of these two NAC filters throughout the 200 days of their operation.

Atrazine biodegradation in bench-scale experiments. In the first experiment, various atrazine media containing 8 μg/l of atrazine were inoculated with bacteria taken from various bed depths (0.1 m, 0.6 m and 1.1 m) of atrazine-spiked pilot plant GAC filters (F1 and F4). The GAC filters were sampled in April 1995 (water temperature 12°C). To detach the bacteria, GAC was ultrasonicated in demineralized water for three minutes (Soniprep 150, amplitude 16 μm). This water was then filtered through a paper filter (pore size > 4.5 μm) to remove carbon particles. Media were inoculated with 1 ml of this filtrate. Microbial activity in the inoculum was confirmed by the standard plate count (> 10^4 CFU/ml). All media contained basic salts listed in Table 4.1 (Feakin *et al.*, 1995) and micronutrients listed in Table 4.2 (Hooidonk, 1995). Carbon (3 mg-C/l) and/or nitrogen (175 mg-N/l) were added to three media in a following way: to one medium only carbon was added, to one medium only nitrogen was added and to one medium both carbon and nitrogen were added. Fourth medium had no additions, while the control fifth medium had no additions and was the only medium that was not inoculated. The mixture of the carbon-containing compounds included 2.5 mg-C/l of oxalate, 0.3 mg-C/l of acetate, 0.1 mg-C/l of glyoxal and 0.1 mg-C/l of formaldehyde. Thus the mixture added was the same as the one spiked to the pilot plant GAC filters, but the concentrations were 10 times higher. Nitrogen was added as NH_4NO_3. Media were incubated for five weeks, in the dark, at 30°C, and were shaken at 175 rpm on a laboratory rotator (G2, New Brunswick Sci.). Oxygen was supplied through cotton plugs. Concentration of atrazine, desethylatrazine and desisopropylatrazine in the media was monitored each week. Microbial activity in the media was determined by the standard plate counts after two and four weeks of incubation.

Table 4.1 Concentration of the basic salts in the media.

Compound	K_2HPO_4	$CaSO_4$	$MgSO_4$*$7H_2O$	$FeSO_4$*$7H_2O$
(mg/l)	200	200	200	1

Table 4.2 Concentration of the microelements in the media.

Compound	H_3BO_3	$MnCl_2$ *$7H_2O$	$ZnSO_4$ *$7H_2O$	$CuSO_4$ *$5H_2O$	$Co(NO_3)_2$	$(NH_4)_6Mo_7O_{24}$ *$4H_2O$
(mg/l)	2.86	1.81	0.22	0.08	0.05	0.02

In the second experiment, media containing basic salts (Table 4.1), micronutrients (Table 4.2) and various concentrations of atrazine (1, 3 and 30 μg/l) were inoculated with bacterial consortium obtained from Ciba-Geigy. These unidentified bacteria were shown to degrade atrazine (Stucki, 1996). The experiment was carried out in order to demonstrate that (at least) some atrazine-degrading bacteria can grow in the media used. Atrazine media were inoculated with 1,5-2 ml of glucose-containing solution in which Ciba-Geigy bacteria have been grown for 3 weeks. The media were incubated for two weeks, in dark, at 30°C, and were shaken at 175 rpm on a laboratory rotator (G2, New Brunswick Sci.). Oxygen was supplied through cotton plugs.

In the third experiment, media containing 10 mM phosphate buffer (pH = 7.5), basic salts (Table 4.1) micronutrients (Table 4.2) and 15 mg/l atrazine was passed at a rate of 2 l/day through two glass columns, each filled with 300 ml of glass bead (EBCT = 3.5 h). The feed bottle was stirred continuously to ensure oxygen saturation (8 mg/l) of the column-influent. One column was inoculated with atrazine degrading bacteria that were obtained from Ciba-Geigy. The column was inoculated with 10 ml of an enrichment culture on atrazine and 10 ml of a suspension of cultures grown on atrazine and glucose. Both columns were operated in parallel, for 21 days and at 30°C. During this time, the concentration of atrazine, desethylatrazine and desisopropylatrazine was monitored in the influents and the effluents of the two columns.

Atrazine adsorbed onto GAC. The two GAC filters (F1 and F4) were continuously spiked with atrazine during three years (1107 days) of their operation. During the third year of operation the concentration of atrazine in the influents and effluents of GAC filters was measured on only two occasions (day 838 and 1107). Atrazine concentration in the influent of these filters was indirectly monitored by regular adjustment (if necessary) of the influent-flow and the flow of atrazine-dosing pumps. After 838 days of operation the flow through the filters started to vary and decreased below 180 l/h because of increased head-loss. In order to preserve the stratification of GAC bed, the filters were not backwashed. The same effluent-flow (136 l/h on average) was set daily for both ozonated and non-ozonated filter.

After three years of pilot plant operation, GAC from the top 35 cm of the filter that received non-ozonated influent (F1) and the filter that received ozonated influent (F4) was taken out. GAC was dried for 19 days at 45°C. This brought the moisture content of GAC down to 6% of the dry weight. The moisture content was determined as a difference in the weight of GAC before and after additional drying for 4 days at 95°C. Twelve representative samples of each batch of GAC were obtained by splitting (each batch being the top 35 cm of the GAC filter bed). Each sample weighed approximately 30 mg. After the GAC was ground, atrazine was extracted from it by 24 hour-long Soxhlet-extraction with ethyl-acetate. The recovery of this method was determined as 95±3%, 83±2%, and 80±2% when atrazine and GAC were in contact for 2 days, 14 days, and 28 days, respectively (Bandjar, 1996).

4.3.3 Biodegradation of BOM and its preloading

In this bench-scale experiment GAC (NORIT ROW 0.8S) was preloaded with ozonated pretreated Rhine River water in two ways: directly and after this water passed through the filter filled with non-activated carbon (Fig. 4.3). NAC particles were of the same shape and size as the GAC particles used. The ozone dose was 1.3 mg O_3/l. The experiment was conducted in parallel at water temperatures of 12°C and 26°C. NAC filters had a bed depth of 1.8 m and were operated with an EBCT of 60 minutes. Ozonated pretreated Rhine River water was passed through for four weeks (at either 12°C or 26°C) before their effluent was directed to GAC filters. This was done to initiate the bioactivity in NAC filters and to saturate the adsorption capacity of NAC. GAC filters had a bed depth of 5 cm and were operated with an EBCT of 1.7 minutes. The inner diameter of both GAC and NAC filters was 5 cm. Before the effluent of the NAC filters was directed to the GAC filters, ultra-pure water (permeate of the reverse osmosis plant) was passed through the GAC filters for 3 days (v = 2.5 m/h). This was done to eliminate problems associated with the potential increase in pH and the precipitation of calcium carbonate onto GAC particles.

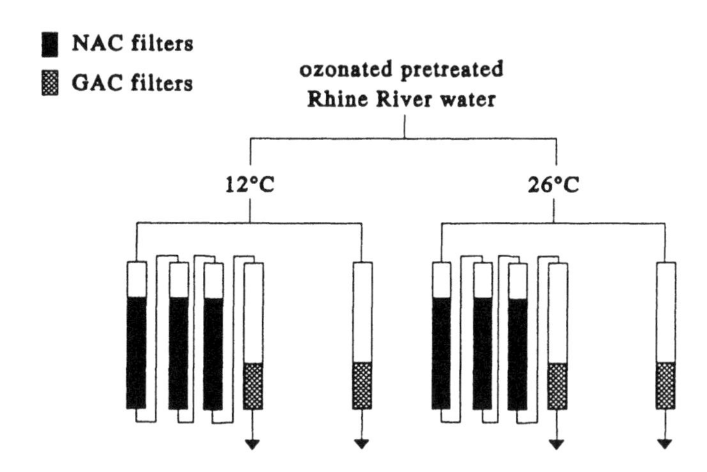

Figure 4.3 Scheme of the experimental set-up used to preload GAC with ozonated pretreated Rhine River water directly and with this water after it passed through filters filled with non-activated carbon.

The preloading of GAC was stopped after two weeks, and demineralized water spiked with 3.5 μg/l of atrazine was passed through the GAC filters. This was done at the same temperature (of about 18°C) for all four filters. The EBCT applied was 1.7 minutes. Atrazine concentration was monitored in the influent and effluent of the four filters for three days.

4.4 RESULTS AND DISCUSSION

4.4.1 Biodegradation of BOM

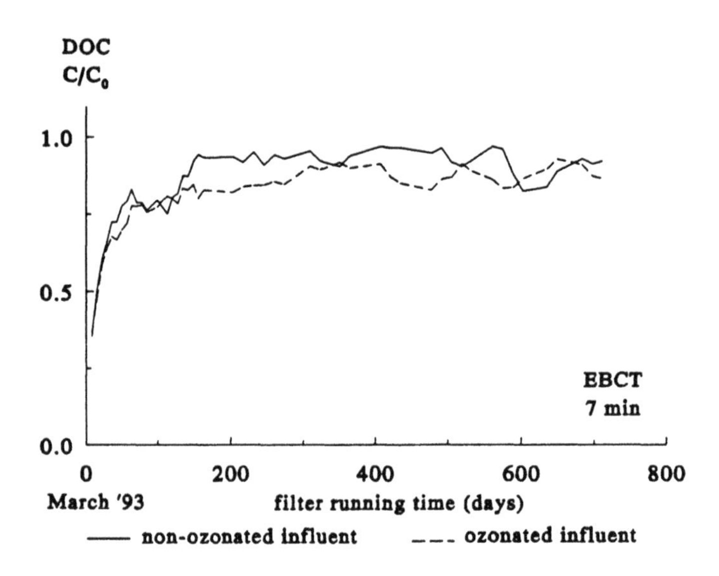

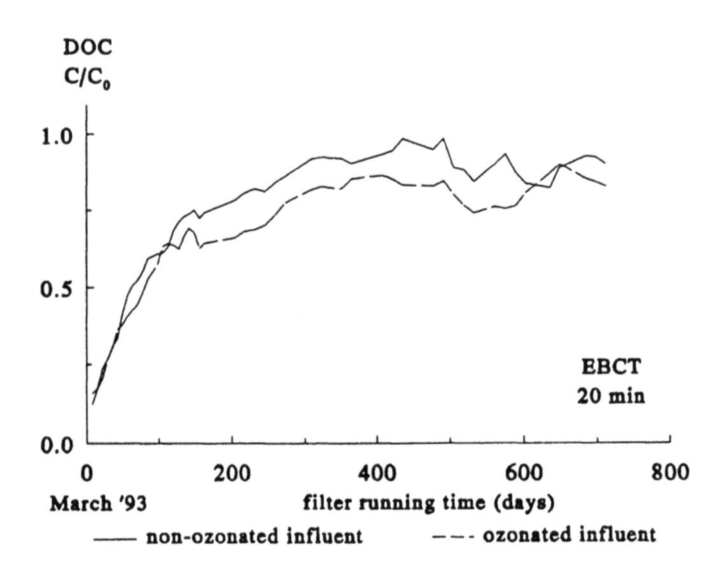

Figure 4.4 DOC breakthrough (3-moving averages) at EBCT of 7 minutes (top graph) and 20 minutes (bottom graph) in GAC filters that received non-ozonated and ozonated pretreated Rhine River water.

Fig. 4.4 shows 3-moving averages for the breakthrough of Dissolved Organic Carbon (DOC) in GAC filters that received ozonated and non-ozonated influent. Measured as DOC concentration, Background Organic Matter was removed better in the GAC filter that received ozonated influent than in its non-ozonated counterpart. Improved BOM removal was observed at both 7 minutes and 20 minutes empty-bed-contact-time (EBCT). The average effluent DOC concentrations of 1.8 mg/l and 1.7 mg/l at an EBCT of 7 minutes, and 1.6 mg/l and 1.4 mg/l at an EBCT of 20 minutes, were observed for the filters that received non-ozonated and ozonated influent, respectively. The average DOC concentration in both non-ozonated and ozonated influent was 2.1 mg/l. No pronounced effect of temperature on the removal of DOC in GAC filters was observed.

Amsterdam Water Supply has observed that aluminum and calcium are released from virgin and regenerated activated carbon grains. This release results in an increase in pH which may cause precipitation of calcium carbonate. In addition the dissociation of BOM is larger at high pH values, which reduces its adsorbability. These effects are not expected to play a role in the aforementioned experiments because the pilot-plant GAC filters were first run for 11 days with the permeate of the reverse osmosis plant (v = 1 m/h). Moreover the run length of two years applied in our experiments is very long compared with the few days that were found required to release calcium and aluminum from GAC (AWS, 1996).

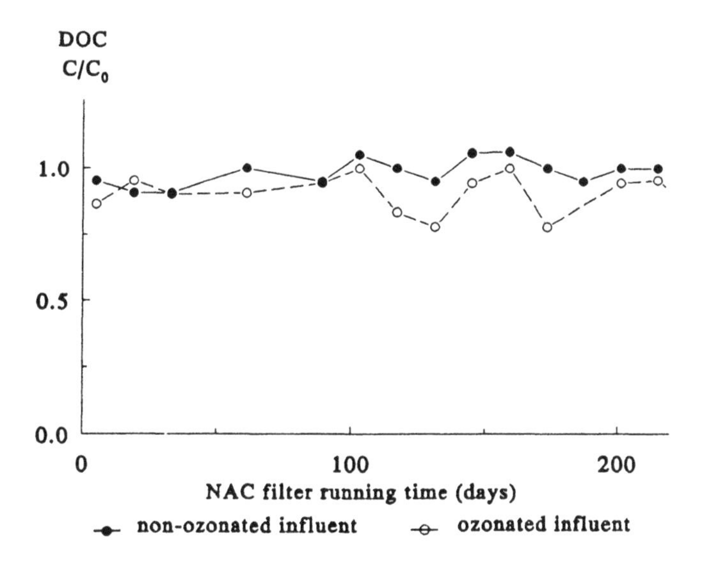

Figure 4.5 DOC breakthrough in non-activated carbon filters (EBCT 20 minutes) that received non-ozonated and ozonated pretreated Rhine River water (DOC = 2 mg/l).

The improved removal of BOM observed in the GAC filters receiving ozonated influent is expected to be due to higher biodegradation of BOM in these filters. Results of the experiment with non-activated carbon filters, in which BOM removal is via biodegradation only, confirmed this expectation (Fig. 4.5). While after a few weeks of operation practically no BOM was removed in the filter that received non-ozonated influent, an average DOC removal of 0.2 mg/l was observed in its ozonated counterpart. The average DOC concentration in both non-ozonated and ozonated influent was 2.0 mg/l.

Higher BOM biodegradation in filters receiving ozonated influent increases the density of bacteria in these filters. These bacteria may, possibly, degrade pesticides and other organic micropollutants as well. In addition, increased BOM removal via biodegradation rather than adsorption can reduce both the competitive adsorption and the preloading of BOM, resulting in better adsorption of organic micropollutants.

4.4.2 Biodegradation of atrazine

Presence of atrazine metabolites. Neither desethylatrazine nor desisopropylatrazine, which are expected as metabolites of atrazine biodegradation, was detected in the effluent of atrazine-spiked GAC filters after 7 and 20 minutes empty-bed-contact-time. Therefore, no indication for atrazine biodegradation in the filters receiving either non-ozonated or ozonated influent was found in this experiment.

Strictly speaking, these results do not exclude biodegradation of atrazine in GAC filters. Namely, it can not be excluded that these two compounds were formed, but that they were present in the effluent of GAC filters only at concentrations that are below the detection limit of the analytical method used (0.03 μg/l). Such low effluent concentrations may have been caused by biodegradation and/or adsorption of these two compounds in GAC filters. In addition, it can not be excluded that atrazine was biodegraded via another pathway in which hydroxyatrazine, which was not monitored, is the initial biodegradation by-product.

Addition of biodegradable organic matter. Rapid clogging of GAC filters was observed when either acetate or the mixture of acetate, oxalate, formaldehyde and glyoxal were added to their influent. To enable filtration velocity of 3.3 m/h, for which the allowed head loss in the filter is 60 cm, the filter that received non-ozonated influent (F2) had to be backwashed every two weeks when 110 μg C/l of acetate was added, and every three weeks when 110 μg Ac-C/l of the mixture of compounds was added (Fig. 4.6). Practically the same clogging occurred in the filter that received ozonated influent spiked with these biodegradable compounds (F3).

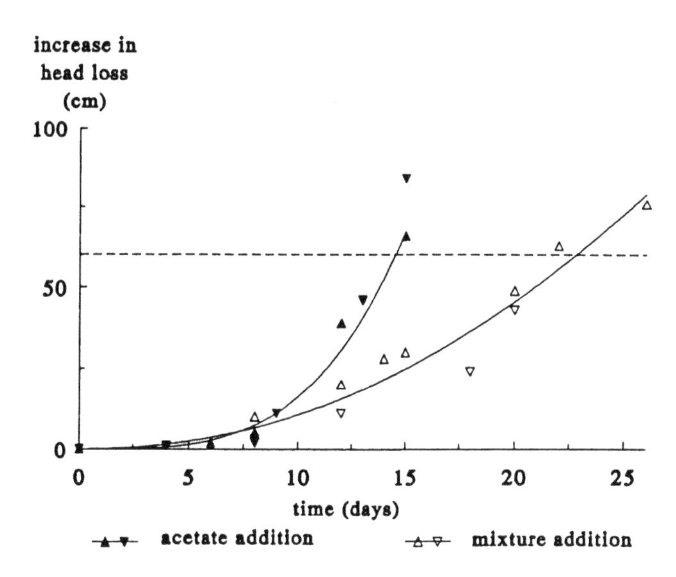

Figure 4.6 Increase in head loss due to the addition of 110 μg C/l of acetate and 110 μg Ac-C/l of a mixture of biodegradable compounds (GAC filter F2, each addition was done twice).

On the other hand, the filters F1 and F4 that received non-ozonated and ozonated influent with the AOC concentration of 8 μg Ac-C/l and 110 μg Ac-C/l, respectively, did not have to be backwashed at all. Frequent backwash of the filters F2 and F3 might have disturbed the stratification of their GAC beds. This made it inappropriate to compare atrazine removal in these filters with the removal in the filters that received the same influent without the addition of biodegradable organic matter. Thus, this experimental approach was abandoned.

The frequent backwashing of the filters spiked with either acetate or the mixture of organic compounds was caused by the intense buildup of biomass at the top of these filters. This can be concluded because practically all head loss occurred within the first 5 cm of the GAC bed. On the other hand, the head loss in the filters that were not spiked was distributed over the full bed depth. These results show that both acetate and the mixture of compounds are biodegraded much faster than ozonated BOM from pretreated Rhine River water. Thus, neither of them describes the composition of ozonated BOM accurately.

Atrazine removal in NAC filters. The results of this experiment did not indicate atrazine biodegradation in Non-Activated Carbon filters receiving either ozonated or non-ozonated influent. Excluding the three points thought to result from an analytical error, the concentration of atrazine in the effluent of NAC filters closely resembled that measured in their influent (Fig. 4.7). Furthermore, neither desethylatrazine nor desisopropylatrazine was detected in any of the 16 samples taken.

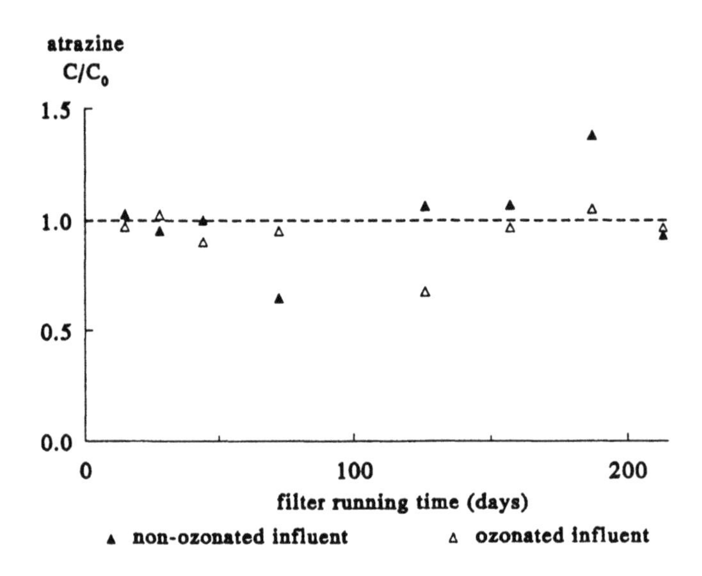

Figure 4.7 Atrazine breakthrough in NAC filters that received non-ozonated and ozonated pretreated Rhine River water spiked with ≈ 3 μg/l of atrazine.

Since these results were obtained for filters filled with non-activated carbon, they do not exclude the possibility that atrazine is biodegraded in filters filled with activated carbon. However, note that so far it has not been shown that a certain micropollutant is biodegraded only in GAC filters. What has been demonstrated is that micropollutants biodegraded in GAC filters are also biodegraded in filters filled with glass beads (Bouwer and McCarty, 1982).

Atrazine biodegradation in bench-scale experiments. Five atrazine-containing media were inoculated with bacteria taken from various depths (*i.e.* 0.1 m, 0.6 m and 1.1 m) of atrazine-spiked pilot plant GAC filters, and incubated for five weeks. Colony counts in these media ranged from 8,6 x 10^5 to 1,4 x 10^6 CFU/ml after two weeks of incubation and from 2,5 x 10^4 to 3,0 x 10^4 after four weeks of incubation. During the five weeks of incubation no reduction of atrazine concentration was observed, and neither desethylatrazine nor desisopropylatrazine was detected in any of the media (Fig. 4.8). Thus, the results did not indicate the presence of atrazine-degrading bacteria in the GAC filters from which the inoculum was taken. However, it can not be excluded that the ultrasonic treatment was not sufficient to remove atrazine-degrading bacteria from the GAC grains or, alternatively, that atrazine-degrading bacteria were inactivated by this treatment. It is also possible that atrazine-degrading bacteria were inactivated due to high salt-concentration in atrazine-containing media or some other difference between these media and the environment in GAC filters. It is however practically impossible to prove these theoretical possibilities.

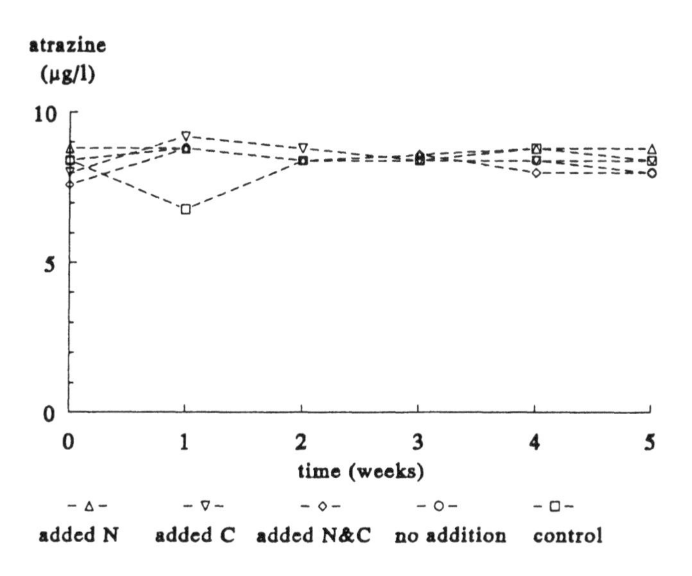

Figure 4.8 Atrazine concentration during the incubation of non-inoculated medium (control) and various media inoculated with microorganisms taken from atrazine-spiked GAC filters.

As indicated by the results of Huang and Banks (1996) who recently demonstrated atrazine biodegradation in lab-scale columns filled with GAC and (as support) glass beads, the aforementioned results do not entirely exclude atrazine biodegradation in GAC filters. The conditions in the pilot-plant GAC filters operated within this research, for which biodegradation of atrazine could not be demonstrated, were however quite different than those in the lab-scale experiment of Huang and Banks. Namely, atrazine concentration in the pilot-plant experiment was 2 µg/l while atrazine concentration in the lab-scale experiment was 50 µg/l. Water temperature in the pilot-plant experiment varied between 2 °C and 21 °C while in the lab-scale experiment it was (presumably) equal to room temperature. Finally, in contrast to the pilot GAC filters, the lab-scale columns were inoculated with bacteria taken from water and sediment of water reservoir with long history of atrazine exposure (to concentrations up to 50 µg/l) and were operated with effluent-recycling (recycling ratio 40:1).

Table 4.3 Atrazine concentration (μg/l) in non-inoculated medium (control) and the three media inoculated with the Ciba-Geigy bacteria at various initial atrazine concentrations.

Sample	control	medium 1	medium 2	medium 3
start	28	25	2.43	0.75
week 1	28	0.56	0.28	0.22
week 2	29	0.22	0.10	0.11

In the second experiment conducted within this part of the research, atrazine-degrading bacteria obtained from Ciba-Geigy were used as the inoculum. As shown by the reduction of atrazine concentration during the two weeks of incubation, these bacteria were degrading atrazine at its initial concentrations ranging from 1 μg/l to 30 μg/l (Table 4.3). Up to 1 μg/l of desethylatrazine and 0.05 μg/l of desisopropylatrazine were also detected in the samples inoculated. These results confirm that atrazine biodegradation may be clearly shown under the experimental conditions applied in this research.

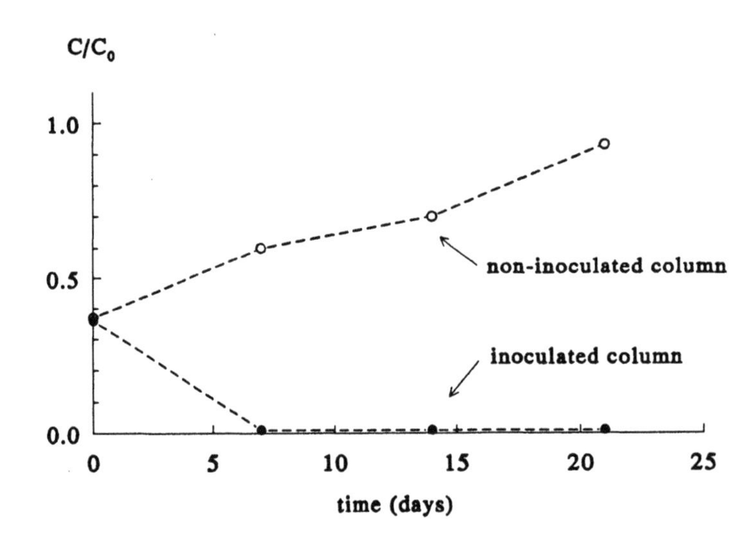

Figure 4.9 Breakthrough of atrazine in the two columns filled with glass beads, one inoculated with atrazine-degrading bacteria and one not inoculated. Atrazine concentration in the influent was 15 mg/l.

Figure 4.9 shows the results of the third experiment in which the concentration of atrazine was monitored in the influents and the effluents of the two columns filled with glass beads, one inoculated with atrazine-degrading bacteria obtained from Ciba-Geigy and one not. At the very start of the experiment (*t*=0) only about 40% of the influent atrazine break through the column. The presumable reason for this was adsorption of atrazine onto glass-beads and/or tubing. Thereafter, as the adsorption capacity of glass-beads and/or tubing was gradually being exhausted, the effluent atrazine concentration in the non-inoculated column was approaching the influent atrazine concentration (100% breakthrough). On the other hand, as clearly demonstrated by practically no breakthrough of atrazine from the inoculated column, the bacteria supplied by Ciba-Geigy were able to grow in this column and to degrade atrazine.

Atrazine adsorbed onto GAC. The mass of GAC in the top 35 cm of the filter that received non-ozonated influent was determined as 7505 g, while the mass of GAC in the top 35 cm of the filter that received ozonated influent was determined as 7438 g. The mass of atrazine removed in the top 35 cm during the initial two years of operation (until day 782) was determined as 3480 mg and 4807 mg in the filter that received non-ozonated and ozonated influent, respectively (see Ch. 2). During the third year of operation (from day 782 until day 1107), the mass of atrazine removed in the top 35 cm of the filter that received non-ozonated and ozonated filter was determined as 562 mg and 983 mg, respectively. The specific (per gram of activated carbon) mass of atrazine removed in the top 35 cm of the filters during the three years of their operation is thus 0.54 mg/g and 0.78 mg/g for the filter that received non-ozonated and ozonated influent, respectively. Therefore, about 44% more atrazine was removed in the filter that received ozonated rather than non-ozonated influent.

The specific mass of atrazine adsorbed onto GAC after three years of pilot plant operation was determined from 12 representative samples of GAC taken from the top 35 cm of each filter (24 samples in total). It was determined as (average $\pm$ σ) 0.64$\pm$0.07 mg/g and 0.96$\pm$0.06 mg/g for the GAC from the filters that received non-ozonated and ozonated influent, respectively. Thus, about 50% more atrazine was found adsorbed onto GAC that received ozonated rather than non-ozonated influent. The expected decrease in the recovery of the desorption method with the prolonged filter running time (Bandjar, 1996) was not accounted for.

These results do not indicate that atrazine is biodegraded in GAC filters or that it is biodegraded better in filters receiving ozonated rather than non-ozonated influent. This would be indicated by the finding that more atrazine is removed from water while less atrazine is found adsorbed onto GAC in the pilot plant filter receiving ozonated rather than non-ozonated influent. However, the results indicate that ozonation of GAC filter influent significantly improves adsorption of atrazine in these filters.

Note that the specific mass of atrazine desorbed from GAC appears to be 0,10$\pm$0,07 mg/g and 0,18$\pm$0,06 mg/g higher than the specific mass of atrazine removed in the GAC filters that received non-ozonated and ozonated influent, respectively. The exact reason for this discrepancy was not found. It is possible that the decreased flow through the GAC filters during the third year of their operation prolonged the actual contact-time in these filters and, as a result, increased the extent of atrazine adsorption. In addition, accounting for the recovery of atrazine analytical method (93%) would increase the difference between the measured influent and effluent atrazine concentrations and would, therefore, increase the calculated mass of atrazine removed in the GAC filters. However, this experiment was done at the very end of the research period and it was not feasible to repeat it and to try to close the mass balance more precisely.

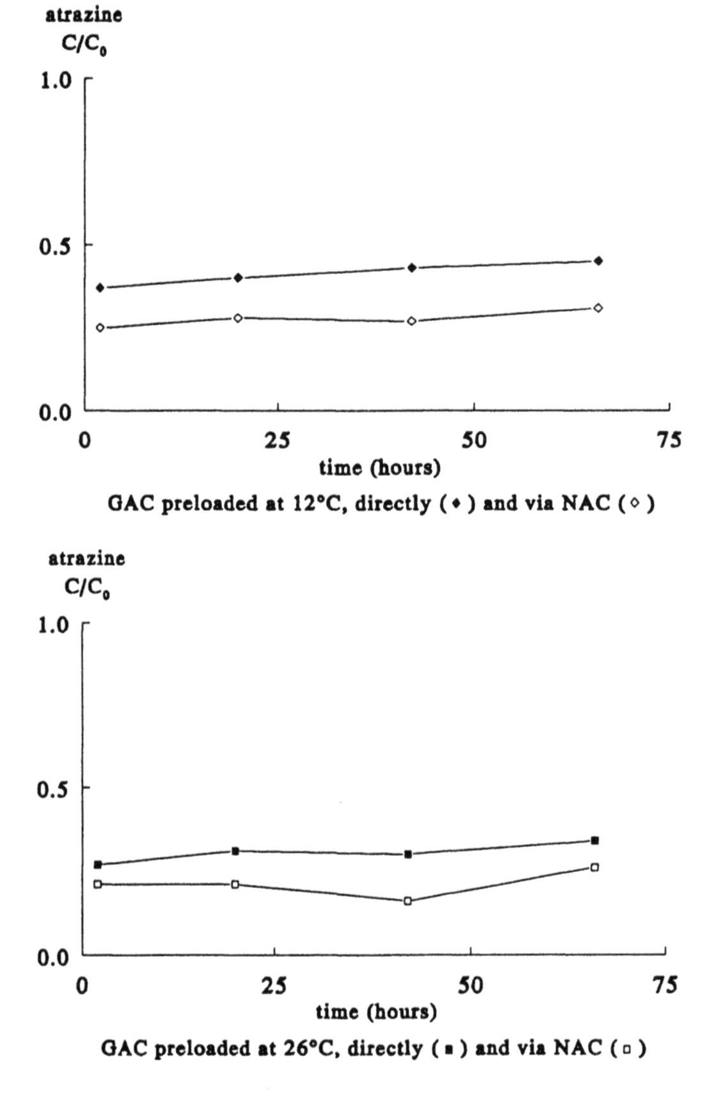

Figure 4.10 Atrazine breakthrough using GAC that was preloaded for two weeks with ozonated pretreated Rhine River water, either directly or after this water passed through Non-Activated Carbon (NAC) filters.

4.4.3 Biodegradation of BOM and its preloading

Less BOM preloading, shown as lower atrazine breakthrough, was observed in short-contact-time (1.7 min) tests with the GAC preloaded with ozonated water that passed through the NAC filters than in tests with the GAC preloaded directly with ozonated water (Fig. 4.10). The observed difference in the average breakthrough percentage was about 33% for GAC preloaded at 12°C, and about 36% for GAC preloaded at 26°C.

The difference in the preloading of BOM can be attributed to biodegradation rather than adsorption of BOM in the NAC filters. This is because these filters were operated for four weeks before their effluent was directed to the GAC filters, which is sufficiently long to saturate the (negligible) adsorption capacity of NAC.

Reduced BOM preloading –and resulting improved adsorption of atrazine– that occurred due to BOM biodegradation in NAC filters is expected to occur due to BOM biodegradation in GAC filters also. Such expectation is based on the assumption that activated carbon does not provide worse support for microbial growth than non-activated carbon.

CONCLUSIONS

Improved removal of BOM observed in GAC filters that received ozonated rather than non-ozonated influent can be attributed to the enhanced biodegradation of BOM in these filters. This can be concluded because an improved removal of ozonated compared with non-ozonated BOM was also observed in NAC filters, in which the removal of BOM is via biodegradation only.

It could not be demonstrated that biodegradation of atrazine accounts for its improved removal in GAC filters that received ozonated rather than non-ozonated influent. No indication of atrazine biodegradation was found in either of the four out of five different experiments conducted (the fifth experiment with the addition of biodegradable organic matter was abandoned because of the intense clogging of GAC filters). Namely, (i) no metabolites of atrazine biodegradation were detected in the effluent of atrazine-spiked GAC filters after 7 and 20 minutes empty-bed-contact-time, (ii) atrazine was not removed in NAC filters, (iii) atrazine was not removed in the liquid media inoculated with the bacteria taken from atrazine-spiked GAC filters, and (iv) after three years of pilot plant operation about 44% more atrazine was removed and about 50% more atrazine was found adsorbed onto GAC in the filter that received ozonated rather than non-ozonated influent.

However, the aforementioned results do not entirely exclude atrazine biodegradation in GAC filters. Atrazine biodegradation was demonstrated during this research for lab-scale columns filled with glass beads while Huang and Banks (1996) demonstrated it for lab-scale columns filled with GAC (to support GAC these columns were also filled with glass beads). Thus, bacteria can be expected to biodegrade atrazine in GAC filters where the same conditions are created. The conditions in the pilot-plant GAC filters operated within this research, for which biodegradation of atrazine could not be demonstrated, were however quite different than those in the aforementioned lab-scale experiments. Namely, atrazine concentration in the pilot-plant experiment was 2 µg/l while atrazine concentration in the lab-scale experiments was 15 mg/l (glass beads experiment) and 50 µg/l (GAC experiment). Furthermore, water temperature in the pilot-plant experiment varied between 2°C and 21°C, while water temperature in the lab-scale experiment with glass beads was constant at a level of 30°C and in the experiment with GAC was (presumably) equal to room temperature. Finally, in contrast to the pilot GAC filters, the lab-scale columns with glass beads were inoculated with atrazine-degrading bacteria and were operated with the addition of basic salts and micronutrients to their influent, while the lab-scale columns with GAC were inoculated with bacteria taken from water and sediment of water reservoir with long history of exposure to atrazine (to concentrations up to 50 µg/l) and were operated with effluent-recycling (recycling ratio 40:1).

The enhanced biodegradation of BOM in the filter that received ozonated influent can explain improved adsorption of atrazine in it. This is concluded because atrazine was found better adsorbed onto GAC preloaded with ozonated water that passed through Non-Activated Carbon (NAC) filters, than onto GAC preloaded directly with ozonated water. Because of negligible adsorption of BOM in NAC filters, only biodegradation of BOM in NAC filters could account for the improved adsorption of atrazine observed in this experiment. However, reduced adsorbability and molecular mass of ozonated BOM can also lower both the competitive adsorption and preloading of BOM in GAC filters. Therefore, they can also contribute to the improved adsorption of atrazine observed in the pilot-plant GAC filter that received ozonated rather than non-ozonated influent.

ACKNOWLEDGMENTS

Gratefully acknowledged are the contributions of many who were involved in this part of the research project. Dr. Gerard Stucki (Ciba-Geigy) provided atrazine-degrading bacteria, and the instructions for their cultivation. Adrian Bandjar, Stéphane Dubray, Denis Foucher, Jennifer Hirst and Beatrix Kiss conducted their B.Sc., M.Sc. or post-M.Sc. studies within this context, trying with much enthusiasm to verify atrazine biodegradation. Gebereselassie Tsegaye, also M.Sc. student of IHE, performed the experiments with BOM preloading. The smooth running of the investigations done at the IHE was guaranteed by the skill and eagerness of the entire laboratory-staff, namely: Fred Kruis, Jouke Hooidonk, Cees Bik, Peter Heerings, Frenk Wiegman and Patricia de Jager.

REFERENCES

Assaf, N.A. and R.F. Turco (1994). Accelerated biodegradation of atrazine by a microbial consortium is possible in culture and soil. *Biodegradation*, 5:29-35.

AWS (1995, 1996). Internal reports.

AWWA Committee (1981). Assessment of microbial activity on GAC. *Journal AWWA*, 8:447-454.

Bandjar, A. (1996). *Extraction of atrazine from granular activated carbon*. M.Sc. thesis, EE-338, IHE, Delft.

Behki, R.M., E. Topp, W.A. Dick and P. Germon (1993). Metabolism of the herbicide atrazine by *Rhodococcus* strains. *App.Env.Microb.*, 59:1955-1959.

Billen, G., P. Servais, P. Bouillot and C. Ventresque (1992). Functioning of biological filters used in drinking water treatment - the Chabrol model. *J. Water SRT - Aqua*, 4:231-241.

Bouwer, E.J. and P.L. McCarty (1982). Removal of trace chlorinated organic compounds by activated carbon and fixed film bacteria. *Env.Sci.Tech.*, 12:836-843.

Brock, T.D. and M.T. Modigan (1988). *Biology of microorganisms.* Prentice Hall International Inc.

Cook, A.M. (1987). Biodegradation of s-triazine xenobiotics. *FEMS Microbiology Reviews*, 46:93-116.

Crawford, J.J., G.K. Sims, R.L. Mulvaney and M. Radosevich (1997). Biodegradation of atrazine under denitrifying conditions. *Appl.Microb.&Biotech.*, 49:5:618-623.

Degrémont (1994). *Removal of micropollutants: the case of pesticides.* Proc. Technical symposium on water treatment, Amsterdam.

Erickson, L.E. and H.L. Kyung (1989). Degradation of atrazine and related s-triazines. *Critical Reviews in Environmental Control*, 1:1-14.

Feakin, S.J., E. Blackburn and R.G. Burns (1995). Inoculation of granular activated carbon in fixed bed with s-triazine-degrading bacteria as water treatment process. *Wat.Res.*, 29:819-825.

Foster, D.M., A.J. Rachwal and S.L. White (1992). Advanced treatment for the removal of atrazine and other pesticides. *Water Supply*, 10:133-146.

Graveland, A. (1994). Application of biological activated carbon filtration at Amsterdam Water Supply. *Water Supply*, 14:233-241.

Graveland, A. and J.P. van der Hoek (1995). Introduction of biological activated carbon filtration at Leiduin. *H_2O*, 19:573-579 (in Dutch).

Graveland, A. (1995). Biological Activated Carbon. In *Proc. INKA Workshop*, IHE, Delft.

Hooidonk, J. (1995). Personal communication.

Huang, C.M. and M.K. Banks (1996). Effect of ozonation on the biodegradability of atrazine in GAC columns. *J.Environ.Sci.Health*, 6:1253-1266.

Huck, P.M., P.M. Fedorak and W.B. Anderson (1991). Formation and removal of assimilable organic carbon during biological treatment. *Journal AWWA*, 12:69-80.

Krasner, S.W., M.J. Sclimenti and B.M. Coffey (1993). Testing biologically active filters for removing aldehydes formed during ozonation. *Journal AWWA*, 5:62-71.

Leeson, A., C.J. Hapeman and D.R. Shelton (1993). Biomineralization of atrazine ozonation products. Application to the development of a pesticide waste disposal system. *Jour.Agr.F.Chem.*, 41:983-987.

Mandelbaum, R.T., L.P. Wackett and D.L. Allan (1993) Mineralization of the s-triazine ring of atrazine by stable bacterial mixed cultures. *App.Env.Microb.*, 6:1695-1701.

Matsumura, F. (1982). Degradation of pesticides in the environment by microorganisms and sunlight. In *Biodegradation of pesticides* (Eds. Matsumura, F. and C.R. Krishna Murti). Plenum Press, New York, p. 67-87.

Miller, G.W. and R.G. Rice (1978). European Water Treatment Practices - The Promise of Biological Activated Carbon. *Civil Engr.*, 48:2:81-83.

Namkung, E. and B.E. Rittmann (1987). Removal of taste- and odor-causing compounds by biofilms grown on humic substances. *Journal AWWA*, 7:107-112.

Olmstead, K.P. and W.J.Jr. Weber (1991). Interactions between microorganisms and activated carbon in water and waste treatment operations. *Chem.Eng.Comm.*, 108:113-125.

Orlandini, E. (1992). *Relating microbial activity to granular activated carbon characteristics.* M.Sc. thesis, IHE, Delft.

Rodman, D.J., A.J. van der Veer and J.F. Holmes (1995). The pre-design of Berenplaat water-treatment works, Rotterdam: additional processes to achieve biologically stable water. *Journal CIWEM,* 4:344-352.

Schottler, S.P. and S.J. Eisenreich (1994). Herbicides in the Great Lakes. *Env.Sci.Tech.*, 12:2228-2232.

Snoeyink, V.L. (1990). Adsorption of organic compounds. In: *Water quality and treatment.* 4th edition, AWWA, McGraw-Hill, p. 781-875.

Sontheimer, H., E. Heilker, M. Jekel, H. Nolte and F. Vollmer (1978). The Mülheim proces. *Journal AWWA*, 7:393-396.

Sontheimer, H., J.C. Crittenden and R.S. Summers (1988). *Activated carbon adsorption for water treatment*, AWWA - DVGW Forschungsstelle Engler Bunte Institut, Karlsruhe.

Stucki, G., C.W. Yu, T. Baumgartner and J.F. Gonzalez-Valero (1995). Microbial atrazine mineralization under carbon limited and denitrifying conditions. *Wat.Res.*, 1:291-296.

Stucki, G. (1996). Personal communication.

Stuyfzand, P.J. and F. Lüers (1997). Behavior of micropollutants during bank-filtration and artificial recharge. *H_2O,* 18:554-562.

Ulrich, M.M, S.R. Muller, H.P. Singer, D.M. Imboden and R.P. Schwarzenbach (1994). Input and dynamic behavior of the organic pollutants tetrachloroethane, atrazine and NTA in a lake. *Env.Sci.Tech.*, 9:1674-1685.

Ying, W. and W.J.Jr. Weber (1979). Bio-physicochemical adsorption model systems for wastewater treatment. *J. Water Pollution Control Fed.*, 11:2661-2677.

Van der Kooij, D. and A. Visser (1976). *Removal of organic compounds in a filter filled with activated and in a filter filled with non-activated carbon and the presence and behavior of bacteria in these filters.* Kiwa report SW-164 (in Dutch).

Van der Kooij, D. (1983). Biological processes in carbon filters. *In: Activated carbon in drinking water technology.* Kiwa/AWWA cooperative research report. AWWA Research foundation.

Van der Kooij, D. and W.A.M. Hijnen (1984). Substrate utilization by an oxalate consuming Spirillum species in relation to its growth in ozonated water. *App.Env.Microb.*, 3:551-559.

Van der Kooij, D. (1990). Assimilable organic carbon (AOC) in drinking water. In: McFeters GA, ed. *Drinking water microbiology: progress and recent developments.* Springer-Verlag.

Wackett, L. (1996). Atrazine biodegradation. *http://dragon.labmed.umn.edu/~lynda/atrazine*

Weinberg, H.S., W.H. Glaze, S.W. Krasner and M.J. Sclimenti (1993). Formation and removal of aldehydes in plants that use ozonation. *Journal AWWA,* 5:72-85.

WHO (1993). *Guidelines for drinking water quality. Vol. 1 Recommendations.* World Health Organization, Geneva, p. 608-614.

Willemse, R.J.N. and J.C. van Dijk (1994). Modeling assimilable organic carbon degradation in biologically active filters. *H_2O,* 2:34-40 (in Dutch).

Yanze-Kontchou, C. and N. Gschwind (1994). Mineralization of the herbicide atrazine as a carbon source by a *Pseudomonas* strain. *App.Env.Microb.*, 60:4297-4302.

Chapter 5

Ozonation and Competitive Adsorption of Background Organic Matter [1]

ABSTRACT—The general aim of the research presented in this chapter is to quantify the extent to which ozonation increases GAC's adsorption capacity for atrazine, due to reduced adsorbability of ozonated Background Organic Matter (BOM). GAC's capacity for atrazine was determined in ozonated and non-ozonated pretreated Rhine River water, both from the experiments and by modeling the competitive adsorption of BOM and atrazine. Experiments were conducted for the range of initial atrazine concentrations that allow accurate performance of an isotherm test ($C_0 > 3$ µg/l). These concentrations are much higher than the concentration of 0.3 µg/l, which is the highest atrazine concentration observed in pretreated Rhine River water in the period 1991-1995. Therefore, the Ideal Adsorbed Solution Theory (IAST) was applied to predict the capacity for C_0 of 0.3 µg/l. The adsorption parameters of BOM required to model its competitive adsorption by the IAST were determined by the Fictive Component Adsorption Analysis (FCAA) method and by the Equivalent Background Compound (EBC) method.
Atrazine adsorption isotherms (C_0 of 3 µg/l and 30 µg/l), determined in ozonated and non-ozonated pretreated Rhine River water, clearly proved that GAC's capacity for atrazine significantly increases due to reduced adsorbability of BOM. No significant difference was found between atrazine isotherms determined for ozone doses from 1 mg/l to 4 mg/l.
The IAST-EBC and the IAST-FCAA methods predict that the equilibrium ($C = 0.1$ µg/l) adsorption capacity for atrazine in non-ozonated pretreated Rhine River water for C_0 of 0.3 µg/l is between 25% and 50% of the capacity obtained from the experiments done for the initial atrazine concentration of 3 µg/l.
The IAST-FCAA method, based on the analysis of DOC isotherms, was found not applicable for the prediction of the adsorption capacity for atrazine in ozonated water. The IAST-EBC method predicts that the equilibrium ($C = 0.1$ µg/l) adsorption capacity for atrazine is about two times higher in ozonated rather than non-ozonated pretreated Rhine River water.

---

[1] Published by E. Orlandini, J.C. Kruithof, J.P. van der Hoek, M.A. Siebel and J.C. Schippers (1996) in *Proceedings of the American Water Works Association Annual Conf.*, C:283-302.

5.1 INTRODUCTION

Significantly improved removal of atrazine (*i.e.* delayed atrazine breakthrough) was observed in GAC filters that received ozonated rather than non-ozonated pretreated Rhine River water (see Ch.2). Ozone-induced oxidation of atrazine did not account for the delayed atrazine breakthrough observed in this experiment, because atrazine was spiked after complete depletion of ozone (O_3 residual < 0.01 mg/l). Thus, improved atrazine removal could result from enhanced biodegradation and/or increased adsorption of atrazine in filters that received ozonated influent. However, it was shown that biodegradation of atrazine, most likely, does not play a significant role in the removal of atrazine in the GAC filters we operated. In contrast, higher adsorption of atrazine in filters receiving ozonated rather than non-ozonated influent was clearly proved. This improved atrazine adsorption may have been caused by reduced competitive adsorption and/or reduced preloading of Background Organic Matter (BOM) in filters receiving ozonated influent (see Ch. 2). The term "BOM" refers to the organic matter in the influent of GAC filters, other than the target compounds that need to be removed. BOM is mostly of natural origin, *e.g.* compounds such as humic substances, but it also includes –especially in Rhine River water– many compounds of anthropogenic origin.

This chapter deals with the competitive adsorption of BOM and atrazine, and the effect ozonation has on it. Competitive adsorption is caused by the presence of many adsorbable compounds in treated water. Because all these compounds compete for the adsorption sites available on activated carbon, the maximum –single solute– adsorption capacity for a target micropollutant (*eg.* atrazine) is decreased. Ozone-induced oxidation of a part of BOM compounds is expected to reduce the competitive adsorption of BOM and atrazine due to two reasons. First, ozonation increases the fraction of BOM that is easily biodegradable and, consequently, it increases the amount of BOM removed in GAC filters via biodegradation rather than adsorption (see Ch. 3). These BOM compounds do not compete with atrazine for the adsorption on GAC. Secondly, ozone-induced oxidation is expected to decrease the adsorbability of a part of BOM compounds with respect to activated carbon (Langlais *et al.*, 1991). Decreased adsorbability reduces the extent to which these BOM compounds can compete for adsorption.

The focus of the research presented in this chapter is on the effect of the reduced adsorbability of ozonated BOM compounds. Thus, the effect that biodegradation of ozonated BOM has on the reduction of the competitive adsorption of BOM and atrazine was not investigated. If ozonation is shown to increase the adsorption capacity for atrazine due to reduced adsorbability of ozonated BOM compounds, this beneficial effect will probably be increased by the biodegradation of ozonated BOM.

The adsorption capacity for a target compound in raw water is commonly determined from the experiments conducted in this water. In these experiments, various masses of activated carbon are allowed to equilibrate at a constant temperature in given volumes of water containing the target compound. The resulting relationship between the mass of a compound adsorbed per unit mass of GAC and the equilibrium concentration of a compound in water is termed adsorption isotherm. When determining an isotherm, it should also be taken into account that the adsorption capacity in multicomponent mixtures depends on the initial concentration (C_0) of the compounds competing (Najm *et al.*, 1991). This poses a problem when these experiments are done at the low initial concentrations of atrazine expected in raw water. Namely, once activated carbon has been added to water, atrazine concentrations fall below the detection limit. Consequently, the adsorption capacity for atrazine that is relevant for the production of drinking water can be determined in two ways. One is to decrease the minimum concentration at which atrazine can be detected (*e.g.* by applying labeled atrazine), another is to predict the competitive adsorption of BOM and atrazine.

In this study, the later approach was adopted. The following specific objectives were set:

- to verify that ozonation increases GAC's adsorption capacity for atrazine due to reduced adsorbability of BOM from pretreated Rhine River water, and to quantify this effect for the lowest initial atrazine concentration at which isotherms can still be accurately determined from the experiments;
- to determine whether the Ideal Adsorbed Solution Theory (IAST) provides acceptable predictions of the competitive adsorption of atrazine and BOM, when BOM's adsorption parameters are determined by the Freundlich method, the Fictive Component Adsorption Analysis (FCAA) method and the Equivalent Background Compound (EBC);
- to predict the adsorption capacity for atrazine in pretreated Rhine River water at the initial atrazine concentrations that are too low to do the experiments, and to quantify the effect of ozonation on this adsorption capacity.

The equilibrium (C = 0.1 μg/l) adsorption capacity of GAC for atrazine in ozonated and non-ozonated pretreated Rhine River water was determined from the experiments for the range of the initial atrazine concentrations that allow them (3-300 μg/l).

For the initial atrazine concentration of 0.3 μg/l, the adsorption capacity of GAC at C of 0.1 μg/l was predicted by modeling the competitive adsorption via the IAST (Radke and Prausnitz, 1972). The concentration of 0.3 μg/l is the highest atrazine concentration measured in pretreated Rhine River water in the period 1991-1995, but is too low to do the experiments. Adsorption parameters of BOM required for the prediction of its competitive adsorption were determined by the aforementioned three methods: the Freundlich method, the FCAA method (Frick and Sontheimer, 1983) and the EBC method (Najm *et al.*, 1991).

5.2 THEORETICAL BACKGROUND

5.2.1 Ideal Adsorbed Solution Theory

The Ideal Adsorbed Solution Theory (IAST) was originally used to calculate multicomponent adsorption of gaseous mixtures (Myers and Prausnitz, 1965), and was later modified by Randtke and Prausnitz (1972) to model the multicomponent adsorption in liquids. The IAST is based on the thermodynamics of adsorption, and relies upon the following assumptions: (i) the same surface area is available to all competing compounds; (ii) there are no adsorbate-adsorbate interactions; (iii) the adsorbent (*e.g.* GAC) is inert, thus, chemisorption does not play a role; and (iv) the multicomponent mixture behaves as an ideal solution.

The last assumption implies the validity of Raoult's law over the entire range of concentrations of a compound. This law states that the relative reduction of the vapor pressure of a dilute solution is equal to the mole fraction of the compound present in that solution. It relates the equilibrium concentration of a compound in water when it is present with other compounds (C), to its equilibrium concentration in water when it is the only compound present (C^o), and to its respective masses adsorbed per unit mass of activated carbon (q and q^o). The IAST uses a two dimensional version of Raoult's law, where temperature (T) and spreading pressure (π) are set equal for the single component and mixture equilibria ($\pi_i = \pi_m$):

$$C_i = C_i^o\,(T,\pi_m)\,\frac{q_i}{\sum\limits_{j=1}^{N} q_j} = z_i\,C_i^o(T,\pi_m) \qquad \textbf{(5.1)}$$

The spreading pressure is the difference between the two surface tensions (σ): one that exists between the adsorbent and pure water, and another that exists between the adsorbent and water containing various compounds. The difference between these two surface tensions is caused by the adsorption of compounds onto adsorbent's surface (A). It represents the work of adsorption that, according to Gibbs, can be related to the amount adsorbed (Sontheimer *et al.*, 1988):

$$-A\,d\sigma = A\,d\pi_i = q_i^o\,R\,T\,d\,ln\,C_i^o \qquad \textbf{(5.2)}$$

Integrating Eq. 5.2 under isothermal conditions, the spreading pressure can be expressed as:

$$\pi_i = \frac{RT}{A}\int_0^{C_i^o} q_i^o \frac{dC_i^o}{C_i^o} \tag{5.3}$$

When the adsorbed phase is modeled as an ideal solution, the mass of a compound adsorbed per unit mass of activated carbon when it is in the mixture of compounds (q) and when it is the only compound present (q^o), can be related according to:

$$\sum_{i=1}^{N} \frac{q_i}{q_i^o} = 1 \qquad (\pi,\ T = \mathit{constant}) \tag{5.4}$$

Equations 5.1, 5.3 and 5.4 are the basic equations of the IAST model.

For a system in which the concentration of a compound in water is reduced only due to its adsorption onto activated carbon, the following mass balance equation can be written (C_0 is the initial compound's concentration, m is the mass of activated carbon added, and V is the volume of a solution):

$$C_0 - C = \frac{m}{V}\, q \tag{5.5}$$

Crittenden *et al.* (1985a) showed that the use of a single Freundlich adsorption-isotherm-equation ($q = K \cdot C^n$) within the IAST model allows successful prediction of competitive adsorption in mixtures of two to six halogenated compounds. More recently, other researchers showed that such an assumption is also valid for the competitive adsorption of atrazine and BOM (Qi *et al.*, 1994; Knappe, 1996). Incorporating the Freundlich equation into the IAST model to express the equilibrium concentration of a compound in water, and combining it with the mass balance equation 5.5, the following equation is obtained (Crittenden *et al.*, 1985a):

$$C_{0,i} - \frac{m}{V} * q_i - \frac{q_i}{\sum_{j=1}^{N} q_j}\left(n_i \frac{\sum_{j=1}^{N} \frac{q_j}{n_j}}{K_i} \right)^{\frac{1}{n_i}} = 0 \qquad i = 1\ \mathit{to}\ N \tag{5.6}$$

Given the initial concentrations of competing compounds and Freundlich coefficients of their single-solute isotherms, equation 5.6 allows prediction of the masses of these compounds adsorbed per unit mass of activated carbon. Since this equation is valid for each of the N compounds present in a mixture, the final equilibrium state can be determined by numerically solving the N nonlinear equations with the N unknown values of q. Once the amount adsorbed is known, the equilibrium concentration of each compound in water (C) can be calculated from the equation 5.5. This procedure is outlined in Figure 5.1, for the case when atrazine and BOM are competing. In this way, the Freundlich adsorption parameters K and n in competing environment can be derived for any initial concentration of competing compounds.

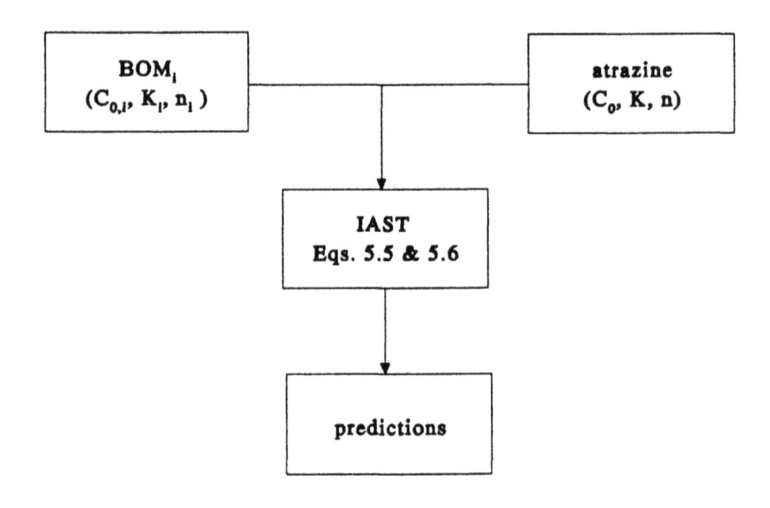

Figure 5.1 Modeling the competitive adsorption of BOM and atrazine by the Ideal Adsorbed Solution Theory (IAST).

5.2.2 Adsorption parameters of BOM

The adsorption parameters of atrazine can be easily determined. Its initial concentration in raw water can be measured, while the coefficients K and n of its single-solute Freundlich isotherm can be deduced from the experiments done in ultra-pure water. The more demanding task is to do the same for the unknown mixture of Background Organic Matter that is present in raw water. In recent years, two general approaches have been developed for that (Fig. 5.2).

One approach is to deduce the adsorption parameters of BOM by analyzing an overall isotherm in raw water. The overall isotherm implies that the measured parameter (*e.g.* DOC, UV) accounts for most of the compounds present in water. Typically, the concentration of

Dissolved Organic Carbon (DOC) is measured for that purpose. The mixture of BOM can be treated either as a single fictive compound, or as two or more fictive compounds.

Another approach is to deduce the adsorption parameters of BOM by analyzing the isotherms of a target compound (*e.g.* atrazine). BOM's adsorption parameters are deduced from the difference between the target compound's single-solute isotherm (*i.e.* its isotherm in ultra-pure water) and its isotherm(s) in raw water. Again, the unknown mixture of BOM can be treated either as a single fictive compound, or as two or more fictive compounds.

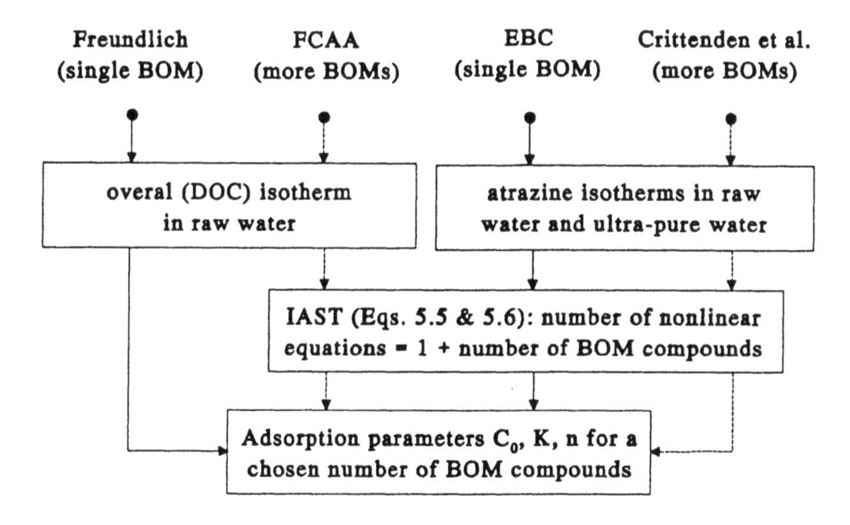

Figure 5.2 Various approaches for the determination of BOM adsorption parameters.

Analysis of DOC isotherms (Freundlich and FCAA method). If a mixture of BOM compounds may be represented with a single fictive compound, the overall DOC isotherm can be treated as a single-solute isotherm. Therefore, it can be fitted by the Freundlich isotherm equation. This results in Freundlich coefficients K and n of the fictive BOM compound. The initial concentration of this fictive BOM compound equals the initial DOC concentration.

The unknown mixture of BOM compounds can also be treated as more than only one compound. This increases the number of unknown adsorption parameters of BOM, and unlike for a single BOM compound, they cannot be determined straight forwardly. The Fictive Component Adsorption Analysis (FCAA) method can be used to determine the adsorption parameters for any, freely chosen, number of fictive BOM compounds. This method was proposed by Frick and Sontheimer in 1983, and it searches for such adsorption parameters

(C_0, K, n) of fictive BOM compounds that minimize the difference between the two DOC isotherms. One of these two isotherms is the DOC isotherm obtained when these adsorption parameters are used to model the competitive adsorption of fictive BOM compounds by the IAST (Eq. 5.5 and Eq. 5.6). The other is the DOC isotherm obtained from the experiments. Commonly, BOM is divided into three fractions: non-adsorbable ($K = 0$), weakly adsorbable (low K), and strongly adsorbable fraction (high K). By further increasing the number of BOM fractions, the accuracy with which a particular DOC isotherm can be fitted is only slightly increased (Sontheimer *et al.*, 1988).

Analysis of atrazine isotherms (EBC method). Compared with atrazine's single-solute isotherm determined in ultra-pure water, atrazine isotherms determined in raw water show lower adsorption capacity (q) for a given concentration of atrazine in water (C). This reduction in the adsorption capacity of GAC for atrazine is caused by the competitive adsorption of BOM from raw water. When the adsorption parameters of BOM are determined by this approach, the following question is posed: what adsorption parameters must fictive BOM compound(s) have in order to cause the same reduction in GAC's capacity as the one observed when atrazine isotherm was determined in raw water?

To answer this question, the single-solute isotherm for atrazine needs to be known, because it gives the adsorption parameters of atrazine. Also, at least one atrazine isotherm in raw water needs to be known. Different adsorption parameters for one or more fictive BOM compound(s) are then tried until the best set of adsorption parameters is found. The best set is the one that minimizes the differences between two atrazine isotherms. One is atrazine isotherm measured in water of interest. The other is atrazine isotherm obtained when that set of adsorption parameters is used by the IAST to predict the competitive adsorption of BOM and atrazine (Eq. 5.5 and Eq. 5.6). Obviously, it is better if atrazine isotherms in raw water have been determined for various initial concentrations of atrazine. Namely, it increases the range of initial concentrations for which the method has been calibrated, and for which the chosen set gives the best prediction.

This approach was first implemented by Crittenden *et al.* (1985b). They modeled the competitive effect of the mixture of five chlorinated aliphatic compounds on the adsorption of chlorodibromomethane, representing this mixture as both one and two fictive compounds.

The Equivalent Background Compound (EBC) method proposed by Najm *et al.* in 1991 can be used to determine the adsorption parameters (C_0, K, n) of a single fictive BOM compound. This method was shown to allow satisfactory predictions of the competitive adsorption of BOM and metazachlor (Najm *et al.*, 1991) and BOM and atrazine (Qi *et al.*, 1994; Knappe, 1996).

5.2.3 Modeling competitive adsorption of BOM and atrazine

The Ideal Adsorbed Solution theory is based on Raoult's law (Eq. 5.1) and, therefore, operates with the molar concentrations of compounds. Consequently, the adsorption parameters of competing compounds need to be expressed in molar units. The recalculation from mass to molar units is simple in case of atrazine, for which the molecular mass is known. However, it is not simple in case of fictive BOM compounds, because their molecular mass is not known.

Thus, when the adsorption parameters of BOM are determined from the analysis of DOC isotherms, modeling of competitive adsorption by the IAST requires the assumption that the molecular mass of BOM and atrazine is the same. This needs to be assumed both when DOC isotherms are analyzed by the Freundlich and by the FCAA method. Under this assumption, the IAST calculations may be done in either molar or mass units. However, because they are determined from DOC isotherms, the adsorption parameters of BOM are in mass-of-carbon units. On the other hand, the adsorption parameters of atrazine determined from its single-solute isotherm are expressed in mass-of-atrazine units. Thus, they have to be converted to mass-of-carbon units. This recalculation is not expected to pose a problem, because the carbon content of atrazine (45%) lies in the range that may be expected for humic substances (Haist-Gulde, 1991). Therefore, the proportion between the mass of BOM and the mass of atrazine is kept the same, whether expressed per mass-of-compound or per mass-of-carbon. This approach was shown to allow prediction of the competitive adsorption between metazachlor (carbon content 60%) and BOM from pretreated Rhine River water (Haist-Gulde, 1991).

The EBC method and the approach developed by Crittenden *et al.* (1985b) do not require the assumption that BOM and atrazine have the same molecular mass. The reason is that they always operate with the same, although assumed, molecular mass of BOM. Namely, these two methods operate with the same molecular mass of BOM both when deducing the adsorption parameters of BOM from atrazine isotherm(s) measured in raw water, and when incorporating these parameters into the IAST to predict atrazine isotherms for low initial concentrations.

5.3 MATERIALS AND METHODS

5.3.1 Adsorption isotherms

Pretreated Rhine River water was ozonated in the pilot plant counter-current ozonation column (height 5 m, diameter 0.1 m), operated at the AWS' Leiduin treatment plant. Two sets of the experiments were done (experiment A and experiment B). Water quality parameters of pretreated Rhine River water during these two sets of experiments are given in Table 5.1.

Table 5.1 Water quality parameters of pretreated Rhine River water during the two experiments.

Experiment	DOC (mg/l)	T (°C)	pH	UV_{254nm} (1/m)
A	2.0	15	7.5	4.7
B	1.7	20	7.5	4.8

In the experiment A, only atrazine isotherms were determined. They were determined in both non-ozonated and ozonated water (ozone doses of 1 mg/l, 2 mg/l and 4 mg/l), at the initial atrazine concentrations of 3 μg/l and 30 μg/l. Isotherms at initial atrazine concentrations lower than 3 μg/l could not be determined, because atrazine concentrations, upon the addition of activated carbon, dropped below the detection limit of the analytical method (≈ 0.03 μg/l).

In the experiment B, both DOC and atrazine isotherms were determined. DOC isotherms were determined in non-ozonated and ozonated (1.1-4.7 mg O_3/l) water. Atrazine isotherms were determined in non-ozonated water, for the initial atrazine concentration of 300 μg/l, and in both non-ozonated and ozonated (4.7 mg O_3/l) water for the initial concentration of 30 μg/l.

The activated carbon used was NORIT ROW 0.8S, pulverized in a way that 75% of particles pass the 75 μm sieve. Pulverized carbon was obtained directly from the manufacturer NORIT NV. All activated carbon fractions obtained were combined and suspended (0.3-5 g AC/l) in demineralized-deionized water. Various volumes (1-20 ml) of these solutions were dosed to bottles, which were then filled with the relevant water to a volume of 1 liter for atrazine isotherms, and to a volume of 300 ml for DOC isotherms. Bottles used for the determination of atrazine isotherms were equilibrated for four days, which proved to be sufficient in order to reach the equilibrium (data not shown). Bottles used for the determination of DOC isotherms were equilibrated for seven days. This time was shown sufficient in order to reach the equilibrium for DOC from pretreated Rhine River water in the study of Haist-Gulde (1991).

Atrazine concentration was measured via an analytical method that involves liquid-liquid extraction with ethyl-acetate, and gas chromatography with nitrogen-phosphorous detection (see Ch. 2). DOC concentration was measured in a continuous flow system, by colorimetric analysis at 550 nm (see Ch. 4).

5.3.2 Determination of BOM adsorption parameters

The adsorption parameters of ozonated and non-ozonated BOM, *i.e.* the initial concentration and the Freundlich coefficients K and n, were determined (i) via Freundlich method, (ii) via the FCAA method, applying the ADSA software-package (Johannsen and Worch, 1994), and (iii) via the EBC method, applying the EBC14 software-package (Knappe, 1996).

Freundlich method. The initial concentration of BOM was assumed to equal the measured DOC concentration. Parameters K and n were determined so that a single Freundlich isotherm describes the measured DOC isotherm as close as possible. Thus, on a log-log scale, a straight line was fitted to the measured points of a given DOC isotherm.

FCAA method. BOM was assumed to consist of three fictive compounds; one non-adsorbable ($K = 0$), one weakly adsorbable, and one strongly adsorbable. The sum concentration of these three fictive components together was assumed to equal the measured DOC concentration. Using the ADSA software-package, the DOC isotherms predicted by the IAST were fitted to the isotherms determined from the experiments. This fitting aimed to minimize the square of the sum of the relative differences between both solid-phase and liquid-phase DOC concentrations in these isotherms (Eq. 5.7). Thus, the method assumed a certain initial set of BOM's adsorption parameters and searched, subsequently, for any new set of parameters for which the sum shown by equation 5.7 would be reduced.

$$min_{FCAA} = \sum_{j=1}^{M} \left(\left| \frac{q_{cj} - q_{mj}}{q_{mj}} \right| + \left| \frac{C_{cj} - C_{mj}}{C_{mj}} \right| \right)^2 \tag{5.7}$$

EBC method. Using the EBC14 software-package, the atrazine isotherms predicted by the IAST were fitted to atrazine isotherms determined from the experiments with raw water. This fitting aimed to minimize the sum of both relative and absolute differences between measured and predicted liquid-phase concentrations of atrazine (Eq. 5.8). Thus, the method assumed a certain initial set of BOM's adsorption parameters and searched, subsequently, for any new set of parameters for which the sum shown by equation 5.8 would be reduced.

$$min_{EBC} = \sum_{j=1}^{M} \left(\left| \frac{C_{cj} - C_{mj}}{C_{mj}} \right| + \left| \frac{C_{cj} - C_{mj}}{C_{M}} \right| \right) \tag{5.8}$$

5.3.3 Prediction of competitive adsorption

Given the adsorption parameters (C_0, K and n) of BOM compound(s) determined by the aforementioned three methods, and the adsorption parameters of atrazine determined from its isotherm in ultra-pure water, the equilibrium solid-phase concentrations of BOM and atrazine in a mixture were calculated by solving the system of IAST equations (Eq. 5.6) via MathCAD 2.0 (Mathsoft Inc., 1987). For the IAST-FCAA approach three equations were set, because there were three competing compounds (*i.e.* atrazine, weakly adsorbable BOM and strongly adsorbable BOM). For the IAST-Freundlich and IAST-EBC approaches, where there were only two competing compounds (*i.e.* atrazine and a single BOM compound), only the following two partial equations (Eqs. 5.9 a & b) needed to be solved. After calculating the equilibrium solid-phase concentrations (q) in this way, the equilibrium concentrations of atrazine remaining in water (C) were calculated by the mass balance equation (Eq. 5.5).

$$C_{0,atr.} - \frac{m}{V} q_{atr.} - \frac{q_{atr.}}{q_{atr.} + q_{BOM}} \left[\frac{n_{atr.}}{K_{atr.}} \left(\frac{q_{atr.}}{n_{atr.}} + \frac{q_{BOM}}{n_{BOM}} \right) \right]^{\frac{1}{n_{atr.}}} = 0 \qquad \textbf{(5.9 a)}$$

$$C_{0,BOM} - \frac{m}{V} q_{BOM} - \frac{q_{BOM}}{q_{atr.} + q_{BOM}} \left[\frac{n_{BOM}}{K_{BOM}} \left(\frac{q_{atr.}}{n_{atr.}} + \frac{q_{BOM}}{n_{BOM}} \right) \right]^{\frac{1}{n_{BOM}}} = 0 \qquad \textbf{(5.9 b)}$$

Before these calculations were done for IAST-Freundlich and IAST-FCAA approach, the adsorption parameters (C_0, K, n) of atrazine were converted from the mass-of-atrazine units to the same mass-of-carbon units in which the adsorption parameters of BOM were expressed. To do this, C_0 was multiplied by 0.45 (45% being the carbon content of atrazine's molecule), K was multiplied with 0.45^{1-n}, while n (dimensionless) remained the same.

5.4 RESULTS & DISCUSSION

5.4.1 Observed adsorption capacity for atrazine

Figure 5.3 shows atrazine isotherms determined in ozonated and non-ozonated pretreated Rhine River water at the initial concentrations of atrazine of 3 μg/l and 30 μg/l (experiment A). At both initial concentrations, the isotherms in ozonated water are significantly (at a 95% confidence level) higher than the isotherms in non-ozonated water. Thus, ozonation significantly improves the adsorption capacity of GAC for atrazine. However, no significant difference was observed between atrazine isotherms obtained for ozone doses between 1 mg/l and 4 mg/l; all of them could be fitted with a single isotherm at a 95% confidence level. Table 5.2 gives the Freundlich coefficients K and n for these isotherms.

Particularly important are the adsorption capacities of GAC obtained for the initial atrazine concentrations expected in raw water. The maximum concentration of atrazine measured in pretreated Rhine River water during 1991-1995 was 0.3 μg/l. This concentration is much lower than 3 μg/l, which is the lowest atrazine concentration for which isotherms could have been determined accurately from the experiments. Thus, there is a need to predict the competitive adsorption of BOM and atrazine.

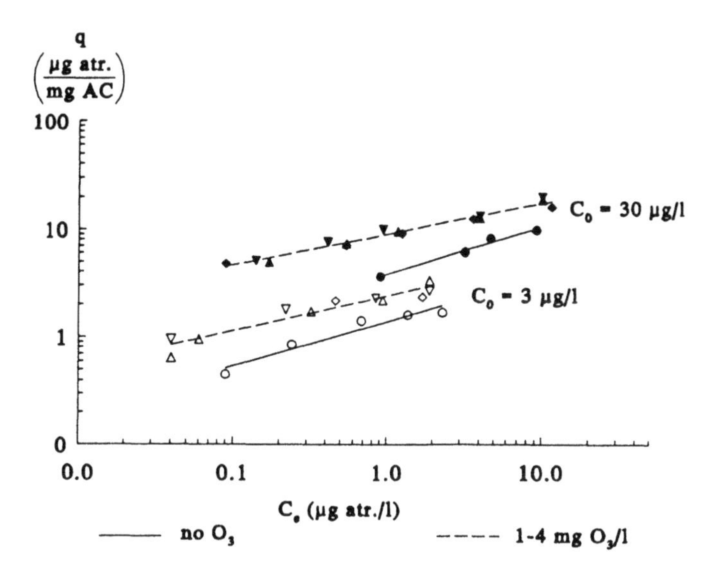

Figure 5.3 Atrazine isotherms in non-ozonated pretreated Rhine River water and pretreated Rhine River water ozonated with ozone doses of 1 mg/l, 2 mg/l and 4 mg/l.

Table 5.2 Freundlich coefficients for atrazine isotherms in pretreated Rhine River water (exp. A).

Freundlich coefficient	non-ozonated water		ozonated water	
	C_0 = 3 μg/l	C_0 = 30 μg/l	C_0 = 3 μg/l	C_0 = 30 μg/l
K (μg/mg)·(l/μg)^n	1.38	3.81	2.39	9.06
n (-)	0.41	0.45	0.32	0.29

5.4.2 Comparison of predicted and observed adsorption capacity

Adsorption parameters of atrazine were determined from its isotherm in demi-water. The Freundlich coefficients K of 41±0.02 (μg/mg AC) (μg/l)$^{-n}$, and n of 0.25±0.01 (average ± standard error) were obtained. The three methods that were applied to determine the adsorption parameters of BOM required either DOC isotherms (Freundlich method and FCAA method) or atrazine isotherms (EBC method) in pretreated Rhine River water. Therefore, both DOC isotherm and atrazine isotherm (C_0 = 300 μg/l) were determined in the same experiment (experiment B). Once determined by these three methods, the adsorption parameters of atrazine and BOM were incorporated into the IAST equations. The predictions of the IAST-

Freundlich, the IAST-FCAA and the IAST-EBC methods were then compared with the atrazine isotherm (C_0 = 30 µg/l) determined during the same experiment. The initial atrazine concentrations of 300 µg/l and 30 µg/l were chosen in order to ensure the accuracy of the determined isotherms.

Freundlich method. Figure 5.4 shows experimentally determined DOC isotherm fitted by Freundlich equation. As shown by this figure, the Freundlich equation provided a reasonable fit of the experimental DOC isotherm. However, the fit resulted in a rather low Freundlich coefficient K of only 0.1 (mg DOC/g AC)(mg DOC/l)$^{-n}$. The reason for this is that Freundlich analysis does not account for the non-adsorbable fraction of BOM. By taking into account the non-adsorbable fraction of BOM, the liquid-phase DOC concentrations would be reduced and, for a given solid-phase DOC concentrations, a higher K will be obtained.

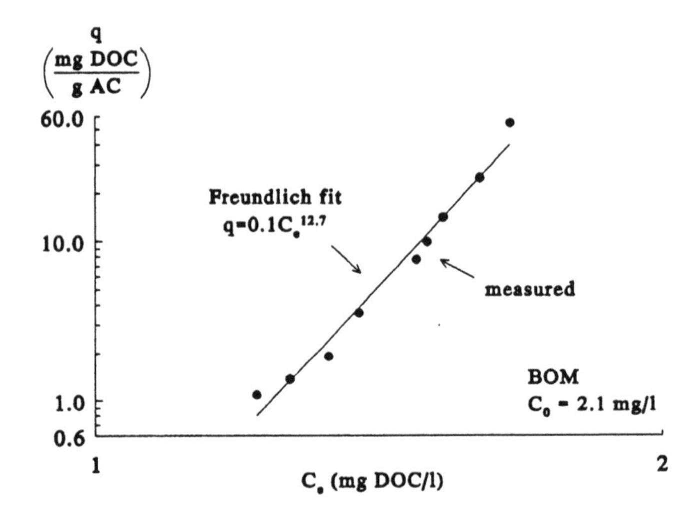

Figure 5.4 Adsorption parameters of BOM in pretreated Rhine River water as determined by the Freundlich method.

FCAA method. Figure 5.5 shows the same experimentally determined DOC isotherm as figure 5.4, and the DOC isotherm obtained using the FCAA method. The concentration of non-adsorbable BOM (1.1 mg DOC/l) was determined by extrapolating the isotherm for the low liquid-phase concentrations (high activated carbon doses). This extrapolation was somewhat tentative, because the region in which increased AC doses do not further decrease liquid-phase DOC concentrations was not reached. Most likely, the AC doses applied were not high enough. The adsorption parameters of the two adsorbable BOM fractions, also shown by this figure, were determined so as to minimize the error (Eq. 5.7 = 0.019) between the experimentally determined DOC isotherm and the one obtained using the FCAA method.

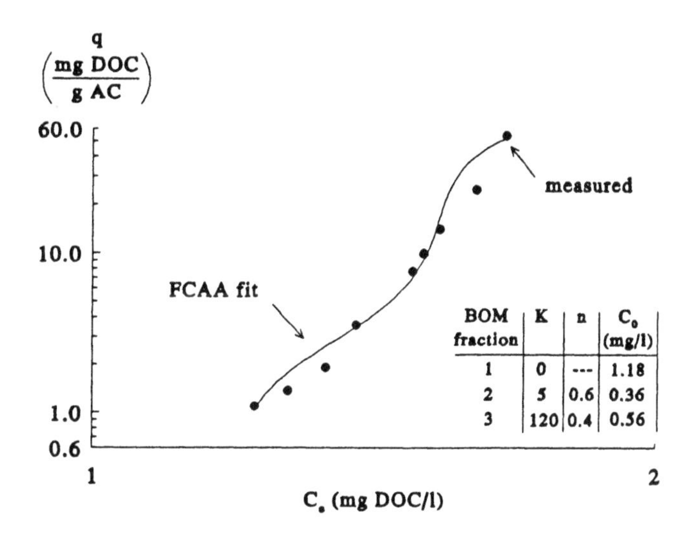

Figure 5.5 Adsorption parameters of BOM in pretreated Rhine River water as determined by the FCAA method.

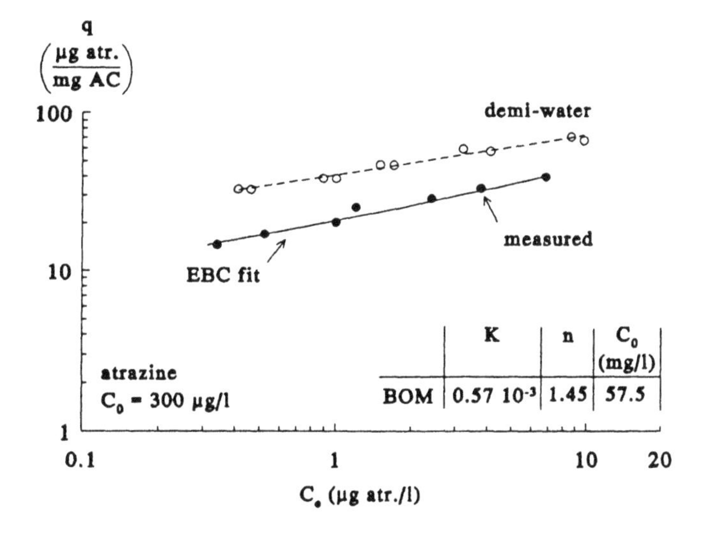

Figure 5.6 Adsorption parameters of BOM in pretreated Rhine River water as determined by the EBC method.

***EBC method*.** Figure 5.6 shows atrazine isotherm in demineralized-deionized water and atrazine isotherm in pretreated Rhine River water (C_0 = 300 μg/l). The error between the isotherm determined from the experiments and the isotherm obtained by the EBC method (Eq. 5.8) was numerically minimized by a somewhat unrealistic set of BOM's adsorption parameters: $K = 0.57 \cdot 10^{-3}$ (mg/g AC)(mg/l)$^{-n}$, $n = 1.45$, $C_0 = 57.5$ mg/l. This type of error can easily happen when the numerical procedure searches for any combination of these three parameters that would minimize the error of the fit. To overcome this problem, it has been suggested to try error-minimization-functions other than the one shown by Eq. 5.8 (Knappe, 1996), or to simplify the IAST equations (Heijman, 1996). The ideal solution, however, will be the numerical procedure that searches for the optimum adsorption parameters only within the defined, physically plausible, boundaries.

***IAST prediction*.** The adsorption parameters of BOM, determined in the above three manners, were incorporated into the IAST model (Eq. 5.5 and Eq. 5.6). The predictions of this model were than compared with atrazine isotherm in pretreated Rhine River water that was determined from the experiments at the initial atrazine concentration of 30 μg/l (Fig. 5.7).

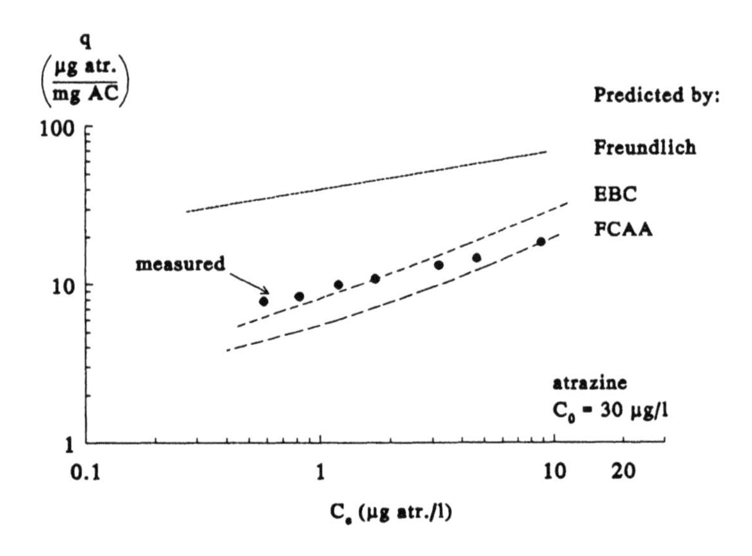

Figure 5.7 Prediction of atrazine isotherm in pretreated Rhine River water by the IAST-Freundlich, IAST-FCAA and IAST-EBC methods.

As shown by this figure, the IAST-FCAA and IAST-EBC methods resulted in reasonable predictions of atrazine isotherm, though the predicted adsorption capacities were somewhat lower for the liquid-phase concentrations that are of practical interest. In contrast to these two methods, atrazine isotherm predicted by the IAST-Freundlich model closely resembled the single-solute atrazine isotherm, suggesting the absence of competitive adsorption between BOM and atrazine. Such a prediction of the IAST-Freundlich method may be expected, due to the rather low coefficient *K* that this method assigned to BOM. Therefore, only the FCAA method and the EBC method were used for further modeling.

5.4.3 Predicted adsorption capacity

In the preceding section, the adsorption parameters of BOM from non-ozonated pretreated Rhine River water were determined. These parameters can be used to predict the adsorption capacity in non-ozonated Rhine River water. To predict the capacity for atrazine in ozonated Rhine River water, the adsorption parameters of BOM from this water need to be determined. This was done by both the FCAA and the EBC method.

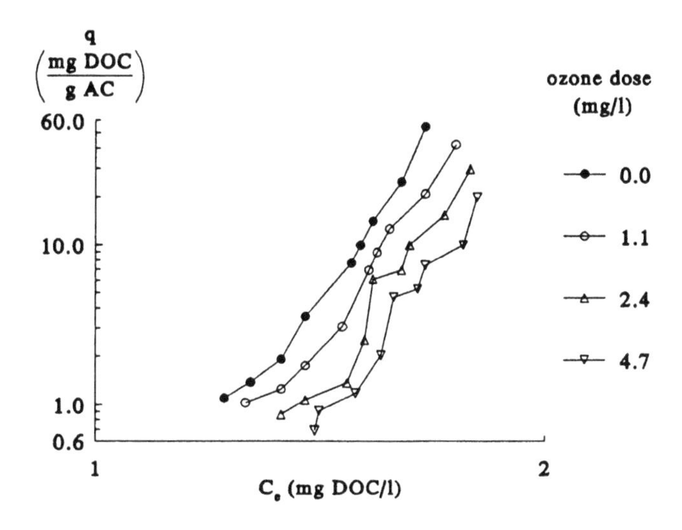

Figure 5.8 DOC isotherms in ozonated and non-ozonated pretreated Rhine River water.

FCAA method. Figure 5.8 gives DOC isotherms determined for non-ozonated and ozonated (1.1- 4.7 mg O_3/l) pretreated Rhine River water. As shown by this figure, ozonation reduces the adsorbability of BOM: there is less DOC adsorbed per unit mass of GAC for a given (equilibrium) DOC concentration in ozonated rather than non-ozonated water. Furthermore, the higher the ozone dose, the lower the adsorbability of BOM. The same effect was also observed in another experiment that was conducted (data not shown). However, surprisingly, this effect of an increased ozone dose was not found reflected in an increased GAC's capacity for atrazine (Fig. 5.3). It is possible that after a certain ozone dose (*e.g.* 1.1 mg/l), ozonation further reduces only the adsorbability of BOM compounds that do not compete with atrazine.

Figure 5.9 shows the results of the FCAA analysis of the DOC isotherms given in figure 5.8. To allow the comparison of BOM adsorption parameters for various ozone doses, the Freundlich coefficients of the two adsorbable BOM fractions were kept constant for the analysis of all DOC isotherms obtained. As shown by this figure, the FCAA method explains the reduction of the adsorbability of BOM as due to the pronounced decrease in the amount of strongly adsorbable BOM.

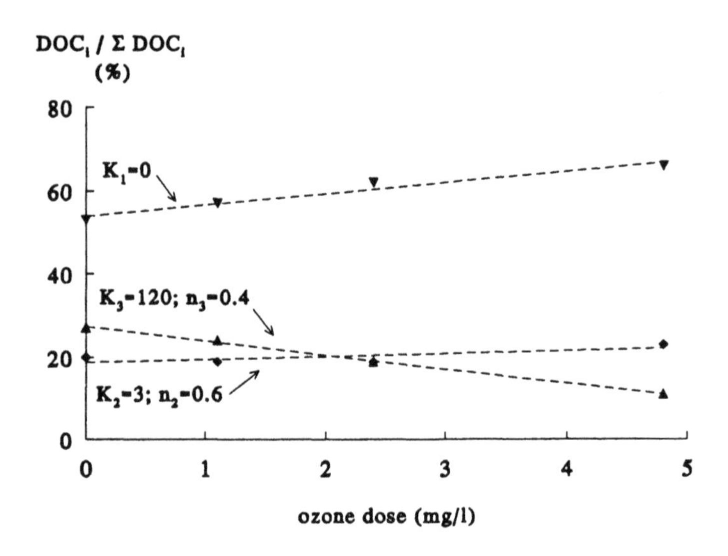

Figure 5.9 Adsorption parameters of BOM in ozonated and non-ozonated pretreated Rhine River water as determined by the FCAA method.

By the FCAA method determined adsorption characteristics of non-ozonated and ozonated BOM were incorporated into the IAST model (Eqs. 5.5 & 5.6) to predict atrazine isotherms at the initial atrazine concentration of 3 μg/l. The predicted atrazine isotherms are shown in figure 5.10. As expected, the FCAA-IAST method predicts that an increase in the ozone dose applied results in a pronounced increase in the adsorption capacity for atrazine; Freundlich coefficient K of 3.1, 4.0 and 6.4 was predicted for ozone doses of 1.1 mg/l, 2.4 mg/l and 4.7 mg/l, respectively (Freundlich coefficient n was 0.7 for all isotherms). Such a prediction is in contrast with the experimental finding that there is no significant difference between atrazine isotherms obtained for various ozone doses (Fig. 5.3). Thus, the FCAA-IAST method does not appear applicable for the prediction of the adsorption capacity for atrazine in ozonated water.

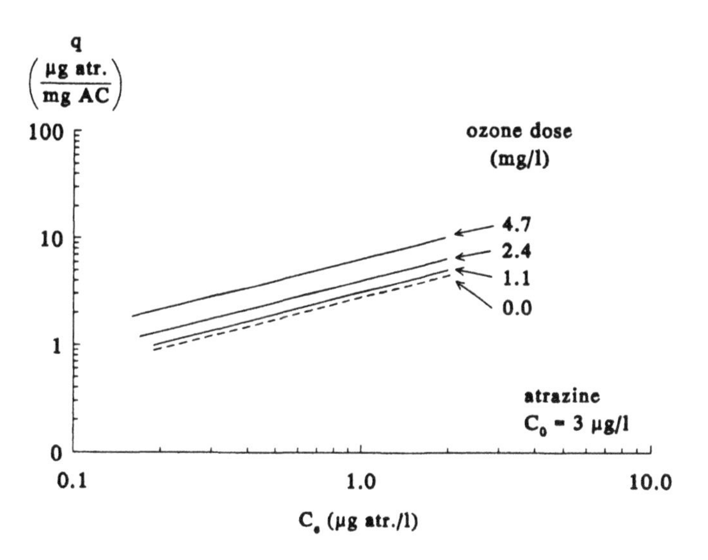

Figure 5.10 By the IAST-FCAA method predicted atrazine isotherms in non-ozonated and ozonated pretreated Rhine River water.

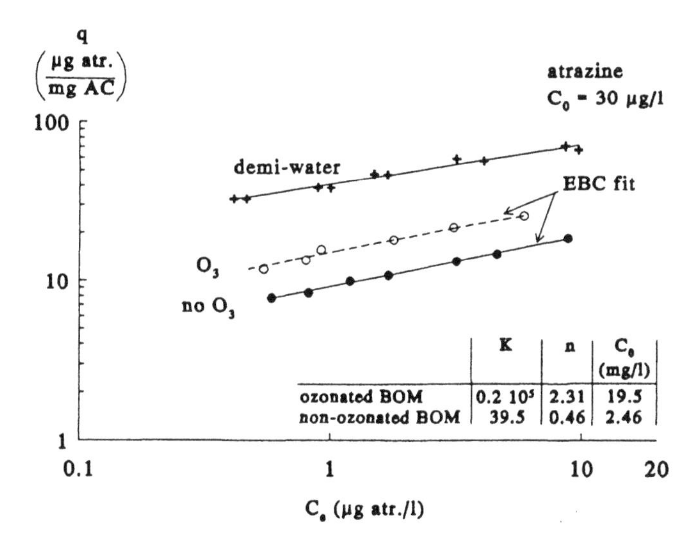

Figure 5.11 Adsorption parameters of BOM in ozonated (4.7 mg O_3/l) and non-ozonated pretreated Rhine River water as determined by the EBC method.

EBC method. Atrazine isotherms in ozonated (4.7 mg O_3/l) and non-ozonated pretreated Rhine River water were used to determine the adsorption parameters of ozonated and non-ozonated BOM. These isotherms were determined for the initial atrazine concentration of 30 μg/l (Fig. 5.11). The adsorption parameters of BOM, determined by the EBC method, are given in the same figure.

IAST prediction. The above determined adsorption parameters of ozonated (4.7 mg O_3/l) and non-ozonated BOM were incorporated, together with the adsorption parameters of atrazine, into the IAST model (Eqs. 5.5 & 5.6). This was done in order to predict the adsorption capacity for atrazine at C of 0.1 μg/l, for various initial atrazine concentrations. Figure 5.12 shows these predictions, and the same adsorption capacity determined from the experiments.

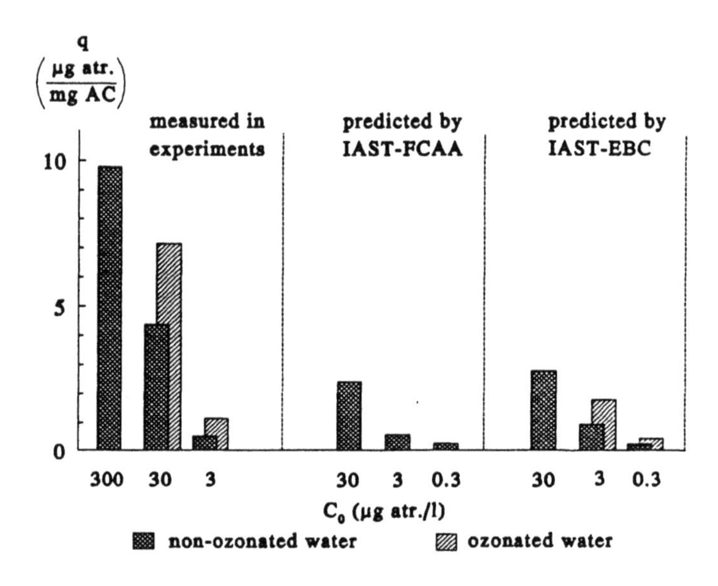

Figure 5.12 Equilibrium (C_e = 0.1μg/l) adsorption capacity for atrazine, measured in the experiments with pretreated Rhine River water and predicted by the IAST-FCAA and IAST-EBC methods.

As shown by this figure, a decrease in the initial atrazine concentration (C_0) reduces the equilibrium (C = 0.1 μg/l) adsorption capacity for atrazine. According to the predictions of the IAST-EBC method and the IAST-FCAA method, this adsorption capacity for C_0 of 0.3 μg/l is between 25% and 50% of the one obtained from the isotherm determined in the experiments done for C_0 of 3 μg/l. Besides, this capacity is only a fraction (2-5%) of the one obtained from the isotherms determined in the experiments done for C_0 between 30 μg/l and 300 μg/l.

The figure 5.12 also shows that ozonation significantly increases the adsorption capacity for atrazine that can be expected in pretreated Rhine River water. The IAST-EBC method predicts that the equilibrium (C = 0.1 μg/l) adsorption capacity for atrazine for C_0 of 0.3 μg/l is about two times higher in ozonated rather than non-ozonated pretreated Rhine River water.

CONCLUSIONS

Ozonation significantly increases the adsorption capacity of GAC for atrazine, due to the reduced adsorbability of ozonated Background Organic Matter (BOM) from pretreated Rhine River water. The same GAC's adsorption capacity for atrazine is obtained for ozone doses between 1 mg/l and 4 mg/l. However, as shown by DOC isotherms, the higher the ozone dose, the lower the adsorbability of BOM compounds.

At the equilibrium atrazine concentration of 0.1 μg/l, the adsorption capacity in ozonated water is roughly doubled compared with the capacity in non-ozonated water. This was clearly shown by the isotherms determined from the experiments done for the initial atrazine concentration of 3 μg/l. This concentration was the lowest atrazine concentration for which isotherms could be accurately determined from the experiments.

The Ideal Adsorbed Solution Theory (IAST), when combined with the Fictive Component Adsorption Analysis (FCAA) method and the Equivalent Background Compound (EBC) method, provides reasonable estimates for the competitive adsorption of atrazine and BOM from non-ozonated pretreated Rhine River water. This is in contrast to the combination of the IAST and the Freundlich method. Such a conclusion was drawn from the experiments done for the initial atrazine concentrations of 30 μg/l and 300 μg/l.

For the non-ozonated pretreated Rhine River water, the IAST-EBC and the IAST-FCAA methods predict that the equilibrium (C = 0.1 μg/l) adsorption capacity for atrazine for the initial atrazine concentration of 0.3 μg/l is between 25% and 50% of the one obtained from the experimental isotherm determined for C_0 of 3 μg/l. Furthermore, this capacity is only a fraction (2% to 5%) of the one obtained from isotherm tests done for C_0 between 30 μg/l and 300 μg/l. This is important, because the concentration of 0.3 μg/l was the highest atrazine concentration measured in pretreated Rhine River water in the period 1991-1995.

For the ozonated pretreated Rhine River water, the IAST-FCAA method –based on the analysis of DOC isotherms– predicts that an increase in the ozone dose applied results in a pronounced increase in the adsorption capacity for atrazine. Such a prediction is in contrast with the experimental finding that there is no significant difference between atrazine isotherms obtained in water ozonated with various ozone doses. Thus, the IAST-EBC method –based on the analysis of atrazine isotherms– can be preferred for the prediction of the adsorption capacity for atrazine in ozonated water. This method predicts that the equilibrium (C = 0.1 μg/l) adsorption capacity for atrazine for C_0 of 0.3 μg/l is about two times higher in ozonated rather than non-ozonated pretreated Rhine River water.

ACKNOWLEDGMENTS

Gratefully acknowledged is the contribution of Pavel Dusilek and Anil Upadhyaya, who conducted their M.Sc. studies within this part of the research project. Pavel's inexhaustible energy and Anils's calm and rational approach were very much appreciated, and provided a sound basis for the results obtained. I also thank Dr. Detlef Knappe (State University of North Carolina) and Dr. Klaus Johannsen (University of Hamburg), for supplying the background information on the software used for the EBC and the FCAA methods, respectively.

SYMBOLS

A = adsorption area per unit mass of adsorbent ($m^2 \cdot kg^{-1}$)
C = concentration of a compound in water in equilibrium with q ($mol \cdot m^{-3}$)
C^o = the same as C, but when only one compound is present in water ($mol \cdot m^{-3}$)
C_{cj} = by IAST calculated C ($mol \cdot m^{-3}$)
C_{mj} = measured C ($mol \cdot m^{-3}$)
C_M = the average (of all compounds) measured C ($mol \cdot m^{-3}$)
C_0 = the initial compound's concentration in water ($mol \cdot m^{-3}$)
K = coefficient of Freundlich's adsorption-isotherm-equation ($mol \cdot kg^{-1}$) ($mol \cdot m^{-3})^{-n}$
m = mass of activated carbon (kg)
M = number of measurements (-)
n = coefficient of Freundlich's adsorption-isotherm-equation (-)
N = number of competing compounds
R = universal gas constant ($J \cdot mol^{-1} \cdot K^{-1}$)
T = absolute temperature (K)
V = volume of solution (m^3)
q = mass of a compound adsorbed per unit mass of activated carbon in equilibrium with compound's concentration C in water ($mol \cdot kg^{-3}$)
q^o = the same as q, but when only one compound is present in water ($mol \cdot kg^{-3}$)
q_{cj} = by IAST calculated q ($mol \cdot kg^{-3}$)
q_{mj} = measured q ($mol \cdot kg^{-3}$)

z = mole fraction of a compound among all the compounds adsorbed (-)

π = spreading pressure ($N \cdot m^{-1}$)

REFERENCES

Crittenden, J.C., P. Luft, D.W. Hand, J.L. Oravitz, S.W. Loper and M. Ari (1985a). Prediction of multicomponent adsorption equilibria using ideal adsorbed solution theory. *Env.Sci.Techn.* 19:11:1037-1043.

Crittenden, J.C., P. Luft and D.W. Hand (1985b). Prediction of multicomponent adsorption equilibria in background mixtures of unknown composition. *Wat.Res.* 19:12:1537-1548.

Frick, B.R. and H. Sontheimer (1983). Adsorption equilibria in multisolute mixtures of known and unknown composition. In M.J.Mc Guire and I.H. Suffet (Ed.) *Treatment of water by granular activated carbon*, Am. Chem. Society, Washington D.C.

Haist-Gulde, B. (1991). *Adsorption of trace-pollutants from surface water*. Ph.D. thesis, University Fridericana, Karlsruhe, Germany.

Heijman, S.G.J. (1996). *Modeling the adsorption of pesticides on activated carbon.* Kiwa Report, SWI 96-241 (in Dutch).

Johannsen, K. and E. Worch (1994). A mathematical method for the performance of the adsorption analysis. *Acta hydrochim. hydrobiol.* 22:5:225-230 (in German).

Knappe, D.R.U. (1996). *Predicting the removal of atrazine by powdered and granular activated carbon*. Ph.D. thesis, University of Illinois, Urbana, USA.

Langlais, B., D.A. Reckhow and D.R. Brink (1991). *Ozone in water treatment*, Lewis Publishers and AWWA Research Foundation.

Mathsoft Inc. (1987). *MathCAD Manual*, Cambridge, USA.

Myers, A.L. and J.M. Prausnitz (1965). Thermodynamics of mixed-gas adsorption. *American Institute of Chemical Engineers Journal.* 1:121-127.

Najm, I.N., V.L. Snoeyink and Y. Richard (1991). Effect of initial concentration of a SOC in natural water on its adsorption by activated carbon. *Journal AWWA*, 8:57-63.

Radke, C.J. and J. M. Prausnitz (1972). Thermodynamics of multi-solute adsorption from dilute liquid solutions. *Amer. Inst. Chem. Eng.*, 18:4:761-768.

Sontheimer, H., J.C. Crittenden and R.S. Summers (1988). *Activated carbon for water treatment,* AWWA - DVGW Forschungsstelle Engler Bunte Institut, Karlsruhe.

Qi, S., S. Adham, V.L. Snoeyink and B.W. Lykins (1994). Prediction and verification of atrazine adsorption by PAC. *J. Env. Eng.,* 120:1:202-218.

Chapter 6

Ozonation and Preloading of Background Organic Matter [1]

ABSTRACT—Preloading of Background Organic Matter (BOM) onto Granular Activated Carbon (GAC) is commonly found to reduce the adsorption of target compounds, such as pesticides. Preloading of BOM is the adsorption of these compounds onto GAC before the adsorption of target compounds. Ozonation of filter influent may reduce this preloading, because it enhances biodegradability, and reduces adsorbability and molecular mass of an important part of BOM compounds.

In this study, the breakthrough of atrazine was determined from the Short Fixed Bed (SFB) tests done with virgin and preloaded GAC. Adsorption capacity of (pulverized) activated carbon for atrazine was determined from adsorption isotherms. Rate-coefficients for atrazine's external and internal mass transfer were determined by fitting the breakthrough of atrazine in the SFB tests with the Plug Flow Homogenous Surface Diffusion model. To allow comparisons, all tests were done in the same ultra-pure water.

The SFB tests confirmed that preloading of BOM from pretreated Rhine River water speeds up the breakthrough of atrazine. Ozonation reduces this adverse effect of preloading: delayed breakthrough of atrazine was observed in the tests with GAC preloaded with ozonated rather than non-ozonated water.

All three parameters that govern the adsorption of atrazine, *i.e.* the adsorption capacity of activated carbon for atrazine, and the external as well as the internal mass transfer rate of atrazine, are reduced due to preloading. Ozonation reduces this adverse effect of preloading: each of these three parameters was found higher for GAC preloaded with ozonated rather than non-ozonated water.

---

[1] Published by E. Orlandini, G.G. Tsegaye, J.C. Kruithof and J.C. Schippers (1997) in *Water Science & Technology*, 7:295-302.

6.1 INTRODUCTION

In drinking water treatment, the influent of GAC filters usually contains many organic compounds. Those that need to be removed by GAC filtration are commonly termed target compounds. Compounds other than the target compounds have been termed either Natural Organic Matter (Sontheimer *et al.*, 1988) or Background Organic Matter (Najm *et al.*, 1991). Considering that many compounds present in Rhine River water are not of natural origin, the term Background Organic Matter (BOM) is used in this research. Adsorption of BOM in GAC filters hinders the adsorption of the target compounds in two ways. They have been termed competitive adsorption of BOM, and preloading of BOM (see Ch. 2). This research focuses on the preloading of BOM (further referred as preloading).

Preloading refers to adsorption of BOM in GAC filters that occurs before the adsorption of target compounds such as atrazine. Preloading is caused by the following. Many BOM compounds are expected to be less adsorbable, of lower diffusivity, and present at much higher concentrations than target micropollutants (*e.g.* atrazine). Because of this, the zone of the filter bed in which these BOM compounds adsorb during a certain time is much larger, and moves faster through the bed than the zone in which micropollutants adsorb. Consequently, deeper in the bed, these BOM compounds adsorb onto GAC before target micropollutants reach there. Preloading has been identified rather recently (in the eighties). It was found to cause a pronounced decrease in both adsorption capacity of activated carbon for target compounds and the mass transfer rates of these compounds on and into GAC. The extent of this decrease is larger than can be explained by competitive adsorption of BOM (Sontheimer *et al.*, 1988).

Significantly improved removal of atrazine (*i.e.* delayed atrazine breakthrough) observed in GAC filters that received ozonated rather than non-ozonated influent (see Ch. 2) may, possibly, be partly explained by lower preloading of ozonated compared with non-ozonated BOM. Namely, ozone-induced oxidation of atrazine did not account for the delayed atrazine breakthrough observed in this experiment, because atrazine was spiked after complete depletion of ozone (O_3 residual < 0.01 mg/l). Thus, improved atrazine removal could result from enhanced biodegradation and/or increased adsorption of atrazine in filters that received ozonated influent. However, it was shown that biodegradation of atrazine, most likely, does not play a significant role in the removal of atrazine in the GAC filters we operated. In contrast, better atrazine adsorption in filters that received ozonated rather than non-ozonated influent was clearly demonstrated. This improved atrazine adsorption may have been caused by reduced competitive adsorption (see Ch. 5) and/or reduced preloading of ozonated BOM. In both mechanisms, enhanced biodegradation of ozonated BOM plays a role (see Ch. 4).

The research presented in this chapter has the following aims:
- to verify that preloading of BOM from pretreated Rhine River water speeds up the breakthrough of atrazine, and that ozonation reduces this adverse impact of preloading;
- to quantify the effect of preloading on the parameters governing adsorption of atrazine (*i.e.* adsorption capacity of GAC for atrazine, and the coefficients of both external and internal mass transfer rate of atrazine), and to quantify the effect of ozonation on these parameters.

Breakthrough of atrazine was determined from the Short Fixed Bed (SFB) tests that were run with virgin GAC and GAC preloaded with BOM from non-ozonated and ozonated pretreated Rhine River water. Adsorption capacity of (pulverized) virgin and preloaded GAC was determined from atrazine adsorption isotherms. The coefficients of atrazine mass transfer rate on/into GAC were searched for by fitting the Plug Flow Homogenous Surface Diffusion (PFHSD) model to the breakthrough of atrazine observed in the SFB tests.

6.2 THEORETICAL BACKGROUND

6.2.1 Preloading and adsorption capacity of GAC

The adsorption capacity of activated carbon is the concentration of a compound adsorbed onto carbon (q) in equilibrium with a certain concentration of that compound in water (C). Commonly, adsorption capacity can be adequately described by simple Freundlich isotherm equation ($q = K \cdot C^n$). The two coefficients of this equation (K, n) are determined experimentally (see Ch. 2). Pulverized rather than intact GAC particles are commonly used in these experiments. Pulverization of GAC shortens the path over which a compound has to diffuse in order to saturate all adsorption sites available onto activated carbon. Consequently, it shortens the time required to reach adsorption equilibrium. This time needs to be shortened, because an unfeasibly long time (*i.e.* months) may be needed to reach the equilibrium with intact GAC particles.

Previous research has shown that the adsorption capacity of (pulverized) preloaded GAC for a target compound is lower than the adsorption capacity of virgin GAC (Zimmer, 1988; Speth and Miltner, 1989; Haist-Gulde, 1991; Carter *et al.*, 1992; Knappe, 1996; Wang and Alben, 1996). Such reduced capacity was observed as a pronounced decrease in the coefficient K, while no clear effect on the coefficient n was observed. Reduced K and the same n suggest that the capacity for adsorption is reduced but, that the distribution of the adsorption energies on the sites available is the same (Weber, 1972).

Studies by Carter *et al.* (1992) and Knappe (1996) showed that, if feasible, adsorption capacity of preloaded GAC needs to be determined in experiments with intact GAC particles. Namely, they found that preloading reduces the adsorption capacity even more than observed in experiments with pulverized GAC: coefficients K for isotherms determined in these two studies for intact particles of preloaded GAC were only 35% to 75% of the coefficients K for isotherms determined for pulverized preloaded GAC. GAC isotherms also had 7% to 20% steeper slopes (*i.e.* higher n) than isotherms determined for pulverized GAC. In the first of these two studies, increased BOM preloading was found to result in a further increase in n.

Carter *et al.* (1992) explained the increased K and reduced n after GAC pulverization as due to the opening of GAC pores that had been physically blocked by preloaded BOM. This may be expected to increase both the number of available adsorption sites (as evidenced by an increased K) and the heterogeneity of their adsorption energies (as evidenced by a reduced n). The later explanation is supported by the work of Weber and Van Vliet (1981). They found that isotherm slopes for adsorption of organic micropollutants on polymeric adsorbents with homogeneous surface characteristics are much higher than those for the adsorption on GACs that have highly heterogeneous surface characteristics.

6.2.2 Preloading and mass transfer rates of target compounds

To get adsorbed onto GAC, the compound first has to reach it. Once brought by the flow of water to the vicinity of a GAC particle (Fig. 2.1, step 1), the compound first needs to diffuse through the stagnant film layer surrounding this particle (Fig. 2.1, step 2). This is termed external mass transfer. Due to the porosity of a GAC particle, the compound can diffuse further inside it. Such internal mass transfer may go either along the surface of GAC pores, or through the water in these pores (Fig. 2.1, step 3). The rate of both external and internal mass transfer of target compounds were found reduced by BOM preloading (Zimmer, 1988; Carter and Weber, 1994; Knappe *et al.*, 1994).

The observed reduction of the external film mass transfer coefficient with preloading may be explained in two ways, as proposed by Carter and Weber (1994). One possible explanation is that the BOM adsorbed on the external surface of GAC creates a zone that has a high resistance to the rate of transport of the target compound. This resistance can be due to the parts of a molecule of BOM that are not adsorbed onto GAC. The same explanation was also proposed by Matsui *et* al. (1994), who successfully modeled the removal of shock-loadings of pesticides in GAC filters by a linear driving force model that assumes that BOM accumulates mainly near the external GAC surface. Another possible explanation for the observed reduction in the external mass transfer rate is the reduction in the effective external surface area of GAC available for atrazine adsorption. The available external surface area may be reduced due to the adsorption of BOM. This reduces the rate of atrazine external mass transfer, because this rate depends on both the film mass transfer coefficient of atrazine and the specific external surface area of GAC (see Eq. 2.2). Thus, when the breakthrough of target compound is modeled assuming that the same external surface area is available on both virgin and preloaded GAC, which is commonly done, the observed reduction in the mass transfer rate on preloaded GAC is explained as due to the decreased film mass transfer coefficient (k_f).

A reduced internal mass transfer rate of a target compound after longer preloading times, expressed either as a reduced pore diffusion coefficient (Carter and Weber, 1994) or as a reduced surface diffusion coefficient (Zimmer, 1988; Knappe *et al.*, 1994), may be explained by increased BOM adsorption in the pores of a GAC particle after processing larger quantities of water. As suggested by Sontheimer *et al.* (1988), the adsorption of BOM hinders the diffusion of the target compound in two ways. The distance over which the target compound needs to diffuse is increased due to the blockage of internal GAC pores by BOM and due to the adsorption of BOM onto the surface of these pores. Also, the target compound may have to counter-diffuse against less adsorbable BOM compounds that have been desorbed. This decreases the rate at which the target compound diffuses.

6.2.3 Exposure time and preloading

All aforementioned studies revealed the same trend regarding the impact of the duration of preloading on the adsorption capacity for target compounds: a rapid decrease during the initial weeks of preloading followed by a more gradual decrease later.

Zimmer (1988) described a reduction of the Freundlich coefficient K as a function of the duration of preloading for many organic micropollutants. This function has an exponential portion for the initial rapid decrease, and a linear portion for the long-term slow decrease. Zimmer determined the coefficients of this function empirically for preloading with Karlsruhe tap water prepared from groundwater. Sontheimer *et al.* (1988) suggested the use of these coefficients for preloading with any groundwater. However, they noted that this approach may not be valid for preloading with surface water, because of the great variation of its quality both in time and space. Consequently, they concluded that the reduction of the adsorption capacity in these cases needs to be determined experimentally for each water type.

The need for this was confirmed by Haist-Gulde (1991) who found that preloading with pretreated Rhine River water resulted in much faster initial decrease in adsorption capacity for trichloroethane than preloading with Karlsruhe tap water. The final capacity loss of about 65% was observed after just four weeks of preloading with Rhine River water, while the capacity loss of 32%, 54% and 60% was observed after 4 weeks, 25 weeks and 34 weeks of preloading with the tap water.

6.2.4 Pretreatment and preloading

There were not many investigations done into the effect of pretreatment on preloading. The only study conducted regarding the conventional treatment of Rhine River water, the one of Summers *et al.* (1989), showed that preloading for four weeks with untreated water and preloading for four weeks with water treated by flocculation and rapid filtration result in the same (40%) decrease of GAC's adsorption capacity for trichloroethane.

Ozonation of GAC filter influent may be expected to reduce the preloading of BOM. First, ozonation transfers a part of BOM in compounds that are more easily biodegradable than the parent compounds. Thus, ozonation increases the concentration of BOM compounds that are removed in GAC filters via biodegradation rather than adsorption. Removal of BOM from GAC filter influent via biodegradation results, as shown in Chapter 4, in less BOM preloading in GAC filters. Secondly, ozonation decreases the adsorbability with respect to activated carbon of a part of BOM compounds (see Ch. 5). Such, less adsorbable BOM, may result in less of a reduction in the adsorption capacity for target compounds. Thirdly, a part of BOM compounds will have reduced molecular mass after ozonation (Langlais *et al.*, 1991). Such smaller BOM compounds can block (partially or completely) the entrance of GAC pores to a lesser extent than the bigger, non-ozonated BOM compounds.

6.2.5 Competitive adsorption and preloading

That BOM and atrazine compete for adsorption during the initial stage of GAC filter operation was shown in the experiments with virgin GAC (see Ch. 5). However, it is possible that after a longer filter running time, BOM has already occupied most of the sites that are attractive for its adsorption. Consequently, when atrazine eventually begins to adsorb onto preloaded GAC, BOM may not compete for the adsorption sites left free.

Making a generally valid conclusion on whether BOM and target compounds compete for the adsorption onto preloaded GAC is not yet possible. If BOM and the target compound compete for the adsorption sites left free, then the adsorption capacity of preloaded GAC in BOM-containing water should depend on the initial concentration of the target compound. In line with this, Haist-Gulde (1991) showed this dependency for metazachlor and GAC preloaded with bank filtered Rhine River water, while Schmidt (1994) showed it for atrazine and GAC preloaded with Hudson River water. On the other hand, no initial concentration dependence and, therefore, no competitive adsorption of BOM was observed for trichloroethane and 1,1,1-trichloroethane when GAC was preloaded with conventionally pretreated Rhine River water (Haist-Gulde, 1991), or for atrazine when GAC was preloaded with Seine River water (Knappe *et al.*, 1994).

The last of these studies is particularly relevant, because it clearly showed that, whatever the extent of competitive adsorption onto preloaded GAC, adsorption capacity of GAC needs to be determined in raw water rather than in organic-free water. While atrazine isotherms determined in this research for Seine River water were independent of the initial atrazine concentration (34 μg/l and 172 μg/l), they show about three times lower adsorption capacity than the isotherm determined for the same preloaded GAC in organic-free water. Given that the BOM surface coverage on GAC after 5 months of preloading was large compared with the number of adsorption sites required for the removal of atrazine at trace levels, and that the GAC used has a large mesopore volume that allows for rapid intraparticle mass transfer, the higher adsorption capacity in organic-free water was explained as due to the desorption of preloaded BOM that freed adsorption sites for atrazine.

6.3 METHODS

6.3.1 Preloading of GAC

GAC (NORIT ROW 0.8S) was preloaded in two pilot plant filters, operated in a down-flow mode with a bed depth of 2.1 m and an empty bed contact time (EBCT) of 20 minutes (Fig. 6.1). Thus, the filtration velocity (assuming an empty bed) was 6.3 m/h, which is within a range of typically applied values (Sontheimer *et al.*, 1988). The influent of GAC filters was Rhine River water pretreated with coagulation, sedimentation and rapid sand filtration (pH of 7.5, temperature from 4.3 to 9.8°C, DOC of 2 mg/l). One GAC filter received non-ozonated influent and one received ozonated influent. During the first two weeks of operation, the ozone dose was 1.9 mg O_3/l, after which it was reduced to 1.3 mg O_3/l. The dose was reduced because of too high ozone residual in the pilot-plant building.

Figure 6.1 The pilot plant used for GAC preloading and G. Tsegaye who conducted his M.Sc. study within this research.

Samples of GAC were taken from 20 cm below the top of the GAC bed, after two weeks and eight weeks of preloading (about 10% and 30% DOC breakthrough in the effluent of the whole filter, respectively). GAC was than dried at 40°C for three days, which was found to bring the moisture content below 1%, and stored in a desiccator until used for the Short Fixed Bed (SFB) tests and the adsorption isotherm tests.

The same DOC removal was observed in both filters: 4.4 mg DOC/g GAC and 17 mg DOC/g GAC after two weeks and after eight weeks of operation, respectively.

Estimating DOC loading onto GAC from these data is difficult due to the following two facts. First, a part of DOC is removed via biodegradation in the filters and does not contribute to DOC loading. Secondly, considering that the filters were not saturated with DOC, the loading on GAC taken from the first 20 cm of a bed depth can be expected to be higher than the average loading over the whole bed depth.

6.3.2 Atrazine breakthrough, adsorption capacity and mass transfer coefficients

All tests were done in deionized-demineralized water (further referred as demi-water) spiked with atrazine. This water was used due to two reasons. First, its use in all experiments allows comparison of the results obtained for GACs preloaded with different water types (*i.e.* ozonated and non-ozonated pretreated Rhine River water). Secondly, the constant quality of this water allows for a valid comparison of the results of the SFB tests conducted at different times.

Atrazine breakthrough. Breakthrough of atrazine was determined in Short Fixed Bed (SFB) tests with virgin and preloaded GAC (Table 6.1).

Table 6.1 Parameters of the Short Fixed Bed (SFB) tests with virgin and preloaded GAC.

Preloading conditions	C_0 (μg atr./l)	mass of GAC (g)	GAC bed depth (cm)	Q (l/h)	v (m/h)	EBCT (sec)
no (virgin GAC)	3.30	28.0	4.2	14.7	7.5	20
non-ozonated water, 2 weeks	2.86	32.6	5.0	11.8	6.0	30
ozonated water, 2 weeks	2.86	32.2	4.9	11.8	6.0	29
non-ozonated water, 8 weeks	2.75	29.9	4.6	11.7	6.0	28
ozonated water, 8 weeks	2.75	29.7	4.6	11.7	6.0	28

These tests were run in down-flow stainless steel columns (inner diameter 5 cm) filled with approximately 30 grams of GAC, which resulted in an immediate atrazine breakthrough. The experiments were conducted at room temperature (20°C). To remove air possibly entrapped while filling the columns with GAC, the columns were backwashed, and demi-water was run through them for one day. Demi-water spiked with ca. 3 µg/l of atrazine was than filtered through the columns for 10 days, while monitoring influent and effluent atrazine concentrations. Biodegradation was assumed not to play a role during these tests, because of the short empty-bed-contact-times applied (24-30 seconds), and the low content of the biodegradable organic matter in demi-water.

Adsorption capacity for atrazine. Equilibrium adsorption capacity for atrazine of virgin and preloaded GAC was determined from Freundlich adsorption isotherms ($q = K \cdot C^n$). GAC was pulverized (75% of particles smaller than 75 µm) to shorten the equilibration time in isotherm tests. This was necessary because 13 weeks of equilibration were still not enough to reach the adsorption equilibrium for atrazine and intact GAC particles (Fig. 6.2). In contrast, 96 hours were shown as sufficient in order to reach the equilibrium with pulverized GAC. No less than 100 mg of pulverized GAC was weighed and suspended in 1 liter of demi-water. Various volumes (3-11 ml) of this suspension were added to bottles containing one liter of atrazine solution. For the isotherm with virgin GAC, the initial concentration of atrazine was 4.3 µg/l. For isotherms with preloaded GAC, the initial atrazine concentration was 2.7 µg/l. Control bottles to which no activated carbon was added were used to measure initial atrazine concentration. The concentration of atrazine remaining in the solution was measured after 96 hours of stirring at 25 °C, and after carbon fines were removed by filtering the solution through pre-rinsed Millipore HV membrane filters with a pore diameter of 0.45 µm.

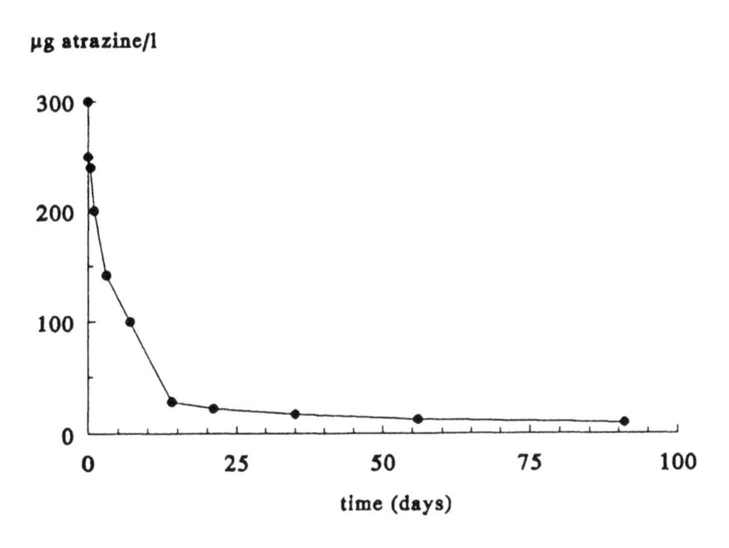

Figure 6.2 Atrazine concentration in water during the equilibration with unpulverized GAC particles.

Atrazine mass transfer. Given the coefficients of atrazine isotherms, external and internal mass transfer rate coefficients of atrazine were determined by fitting the Plug Flow Homogenous Surface Diffusion (PFHSD) model to the breakthrough of atrazine in the SFB tests with virgin and preloaded GAC. This model assumes external and internal mass transfer via film diffusion (Eq. 2.2) and surface diffusion (Eq. 2.3), respectively. The PFHSD-model was fitted using the program written by Yuasa (1992) and modified by Matsui (1994). This program uses a nonlinear least-squares optimization technique to minimize the error between the experimental data and the model output. The same specific external surface area (10.05 m^2/g), equivalent particle diameter (1.04 mm), and apparent and bed density (380 g/l and 334 g/l, respectively) were used for both virgin GAC and preloaded GAC.

6.4 RESULTS AND DISCUSSION

6.4.1 Atrazine breakthrough

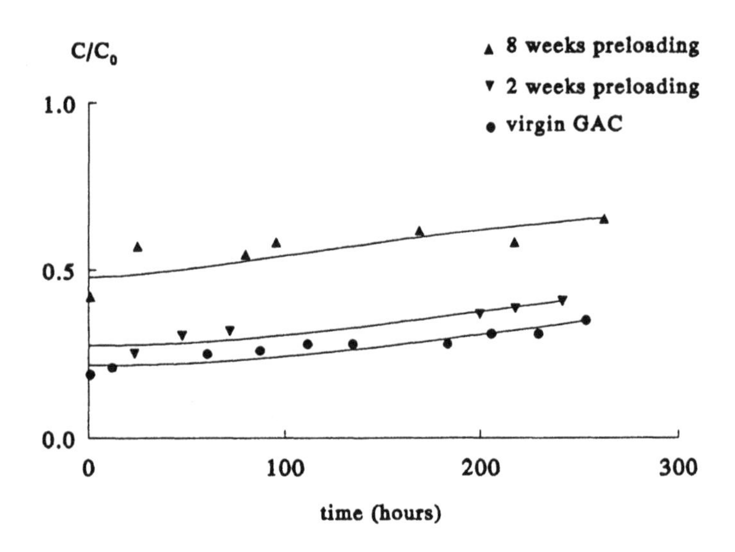

Figure 6.3 Breakthrough of atrazine in the SFB tests with virgin GAC and GAC preloaded with non-ozonated pretreated River Rhine water. Also shown is the fit of the PFHSD model.

A comparison of the breakthrough of atrazine in SFB tests with virgin GAC and with preloaded GAC gives a direct relationship between preloading and the removal of atrazine in GAC filters. This is because these tests are conducted under same flow conditions as those in full-scale GAC filters, and because they do not require pulverization of GAC particles. The later is important because GAC pulverization, which is commonly applied when determining adsorption isotherms tests, was found to increase the adsorption capacity of preloaded GAC (Carter *et al.*, 1992; Knappe, 1996).

Preloading with BOM from pretreated Rhine River water was shown to speed up the breakthrough of atrazine (Fig. 6.3). The effect was found more pronounced for longer preloading times. The breakthrough of atrazine in the SFB test with the GAC preloaded for two weeks was –on average– about 25% higher than the breakthrough of atrazine in the test with the virgin GAC, while this difference increased to 100% after eight weeks of preloading. The fit of the PFHSD model to the breakthrough data resulted in atypical breakthrough curves that were slightly concave-up rather than concave-down. The concave-up shape can be explained as due to low atrazine breakthroughs (20-60%) observed in the SFB tests.

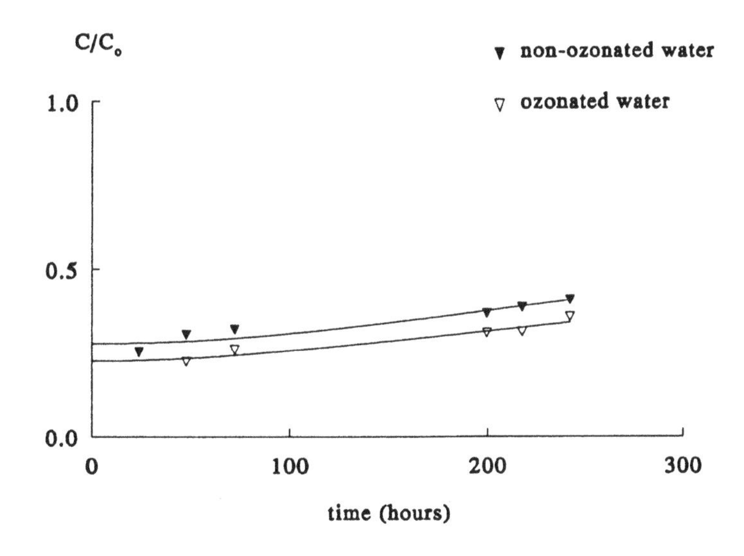

Figure 6.4 Breakthrough of atrazine in the SFB tests with GAC preloaded for two weeks with either ozonated or non-ozonated pretreated Rhine River water. Also shown is the fit of the PFHSDmodel.

Ozonation of GAC filter influent was found to reduce the adverse effect of preloading on the removal of atrazine. Atrazine breakthrough in the SFB tests with the GAC preloaded with ozonated influent was about 17% lower after two weeks of preloading (Fig. 6.4), and about 26% lower after eight weeks of preloading (Fig. 6.5), than the breakthrough of atrazine in the SFB tests with the GAC preloaded with non-ozonated influent for the same period.

These results suggest that after eight weeks of preloading the difference between the GAC receiving ozonated and the GAC receiving non-ozonated influent is higher than after two weeks of preloading. Such a finding does not confirm the expectation that an increased preloading time diminishes the differences between preloading with various BOM types (Haist-Gulde, 1991). A possible explanation is that during the first two weeks of operation, lower preloading in ozonated filter was due to reduced adsorbability of ozonated BOM only, while afterwards increased biodegradation of ozonated BOM started to play a role as well. Another explanation, however, is that the first eight weeks are still the initial phase of preloading, during which both adsorption capacity of GAC and the mass transfer rate of atrazine are rapidly decreasing. In this phase, the difference between the preloading with various water types is expected to increase (Haist-Gulde, 1991).

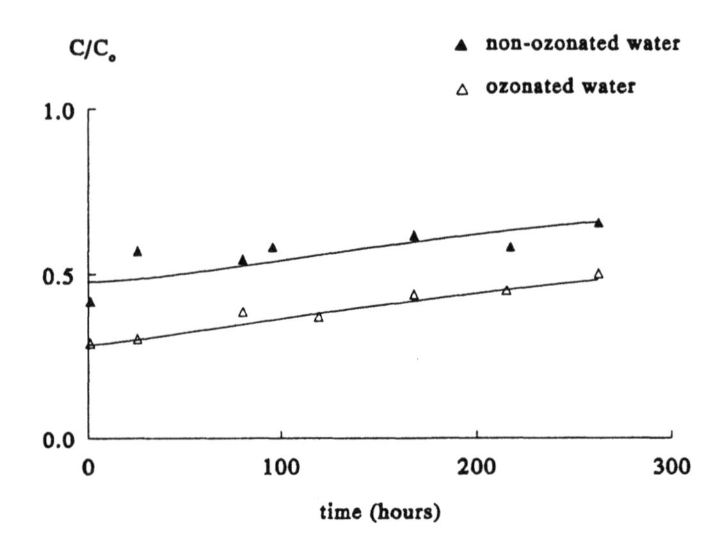

Figure 6.5 Breakthrough of atrazine in the SFB tests with GAC preloaded for 8 weeks with either ozonated or non-ozonated pretreated Rhine River water. Also shown is the fit of the PFHSD model.

Although SFB tests provide a direct relationship between preloading and the breakthrough of atrazine in GAC filters, they do not reveal whether preloading reduces only the equilibrium adsorption capacity of GAC for atrazine, or it also reduces the rate of atrazine transfer on/into GAC. They also do not reveal which of the above is affected by ozonation of GAC filter influent. Knowledge of this helps arrive at a better understanding of the mechanism of BOM preloading and the effect of ozonation on it. Furthermore, it defines adsorption parameters that need to be determined for preloaded GAC in order to model the adsorption of atrazine in GAC filters.

6.4.2 Adsorption capacity for atrazine

Atrazine isotherms for virgin GAC and GAC preloaded with non-ozonated water for two weeks and for eight weeks (Fig. 6.6) showed that preloading with BOM from pretreated Rhine River water reduces the adsorption capacity of GAC for atrazine. This effect was more pronounced for a longer duration of preloading. Comparison of atrazine isotherms for the GAC preloaded with ozonated and non-ozonated water for two weeks and for eight weeks showed that ozonation reduces the extent to which preloading decreases the adsorption capacity of GAC for atrazine (Fig. 6.7 and 6.8). This may be explained as due to enhanced biodegradability, reduced adsorbability and reduced molecular mass, of a sizable portion of BOM compounds that are partially oxidized by ozonation.

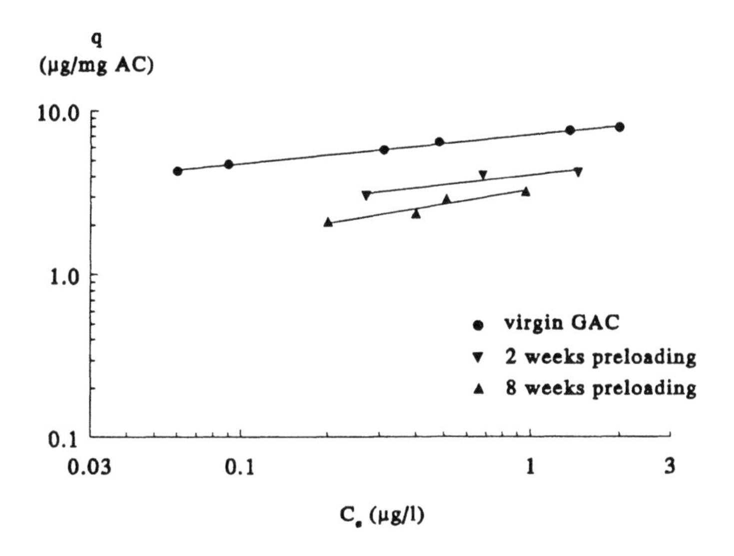

Figure 6.6 Atrazine isotherms for virgin GAC and GAC preloaded with non-ozonated pretreated Rhine River water; GAC pulverized prior to isotherm determination.

Compared with the value obtained for virgin GAC, the value of Freundlich coefficient K for the GAC preloaded with non-ozonated water was reduced for about 43% and 54% after two weeks and eight weeks of preloading, respectively (Table 6.2). The GAC preloaded with ozonated water resulted in a higher value of Freundlich coefficient K than the GAC preloaded for the same time with non-ozonated water. After two weeks of preloading, the difference between the two K values was 19 %, and after eight weeks of preloading it was 32%.

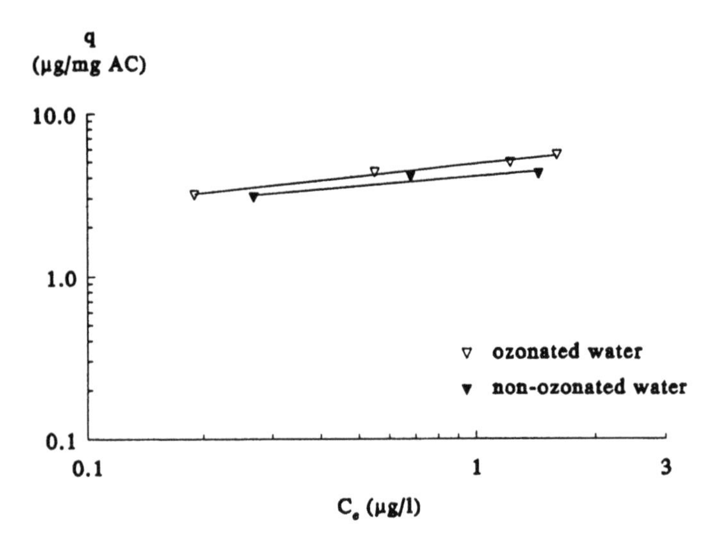

Figure 6.7 Atrazine isotherms for GAC preloaded for two weeks with either ozonated or non-ozonated pretreated Rhine River water; GAC pulverized prior to isotherm determination.

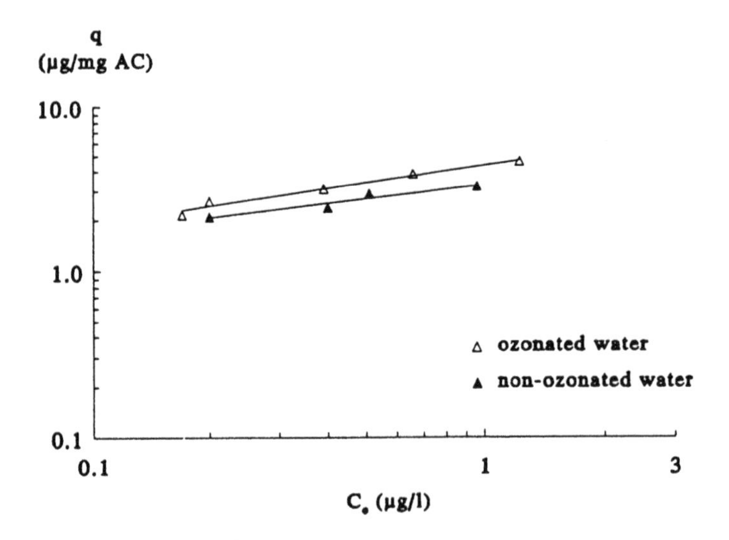

Figure 6.8 Atrazine isotherms for GAC preloaded for eight weeks with either ozonated or non-ozonated pretreated Rhine River water; GAC pulverized prior to isotherm determination.

Table 6.2 Freundlich parameters (K,n) of atrazine adsorption isotherms, and the coefficients of the external (k_f) and internal (D_S) mass transfer of atrazine, for virgin and preloaded GAC.

Preloading conditions	K (mg/g) / (l/µg)	n (-)	k_f (10^{-4} cm/s)	D_S (10^{-13} cm^2/s)
no (virgin GAC)	7.14	0.18	15.0	9.6
non-ozonated water, 2 weeks	4.06	0.20	8.4	9.1
ozonated water, 2 weeks	4.84	0.25	10.0	9.1
non-ozonated water, 8 weeks	3.42	0.29	5.3	4.6
ozonated water, 8 weeks	4.34	0.38	8.8	6.6

The results also indicate that BOM preloading increases the slope n of atrazine isotherms: while the slope for the virgin GAC was 0.18, for preloaded GACs it ranged from 0.2 to 0.36. The slopes of the isotherms obtained for virgin GAC and for preloaded GAC were different at the 95% significance level; however, at the 99% significance level, all the isotherms could have been fitted with 0.3 as a slope. The increased slope of the isotherms determined for preloaded GAC may be explained, as suggested by Carter *et al.* (1992), by preferential BOM adsorption on sites with high adsorption energies. Such BOM adsorption increases the homogeneity of the adsorption energies of the activated carbon sites that are left free for atrazine adsorption. This is expected to result in a steeper slope (higher n) of atrazine isotherms.

6.4.3 Atrazine mass transfer

The results obtained by fitting the Plug Flow Homogenous Surface Diffusion (PFHSD) model to the breakthrough of atrazine in SFB tests show that preloading reduces both external and internal mass transfer coefficients of atrazine (Table 6.2). Thus, the faster breakthrough of atrazine in preloaded compared with virgin GAC filters cannot be fully explained by the reduction of the adsorption capacity of GAC. The external mass transfer coefficient of atrazine obtained after two weeks of preloading was 55% of the one obtained for virgin GAC, whereas the coefficient obtained after eight weeks of preloading was 35% of the one obtained for virgin GAC. For the internal mass transfer coefficient these percentages were 95% and 45%.

Ozonation was found to reduce this adverse effect of preloading: higher mass transfer coefficients were found for GAC preloaded with ozonated than for GAC preloaded with non-ozonated water. The external mass transfer coefficient of atrazine for GAC preloaded with ozonated rather than non-ozonated water was about 20% and 60% higher after two weeks and after eight weeks of preloading, respectively. The internal mass transfer coefficient was the same when GAC was preloaded for two weeks. Possibly, the adsorption of BOM inside GAC pores was not sufficiently high after two weeks of preloading to observe differences between ozonated and non-ozonated water. However, when GAC was preloaded for eight weeks, this coefficient was 45% higher in case of preloading with ozonated rather than non-ozonated water.

The higher external mass transfer coefficient of atrazine for GAC preloaded with ozonated rather than non-ozonated water may be explained in the following way. First, due to the reduced molecular mass of an important part of BOM compounds after ozonation, the layer formed by them around GAC particle may be expected to allow faster diffusion of atrazine than the layer formed by non-ozonated BOM. Secondly, more biodegradable and less adsorbable ozonated BOM compounds may be expected to result in lower surface coverage than non-ozonated BOM. Similarly, the higher internal mass transfer coefficient of atrazine for GAC preloaded with ozonated rather than non-ozonated water may also be expected. Namely, by ozone oxidized BOM compounds are more biodegradable, less adsorbable and of a reduced molecular mass than the parent compounds. Thus, they can result in less coverage of the surface of GAC pores and less blocking of these pores.

CONCLUSIONS

Preloading of Background Organic Matter (BOM) from pretreated Rhine River water speeds up the breakthrough of atrazine in Granular Activated Carbon (GAC) filters. This is in keeping with other studies conducted for many adsorbates and raw water sources. The effect is more pronounced for longer preloading times. Ozonation reduces this adverse effect of preloading; the breakthrough of atrazine for GAC preloaded with ozonated rather than non-ozonated water was 17% and 26% lower after two weeks and after eight weeks of preloading, respectively.

The faster breakthrough of atrazine in preloaded GAC filters rather than virgin GAC filters is due to the reduction in all three parameters that govern the adsorption of atrazine in GAC filters. These three parameters are the adsorption capacity of GAC for atrazine, and the external as well as the internal mass transfer rate of atrazine. Ozonation reduces this adverse effect of preloading. GAC preloaded for two weeks with ozonated rather than non-ozonated water was found to have 19% higher adsorption capacity for atrazine (measured as Freundlich coefficient K), and 20% higher external mass transfer coefficient for atrazine. After eight weeks of preloading, GAC's adsorption capacity, the external mass transfer coefficient of atrazine and the internal mass transfer coefficient of atrazine were 32%, 60% and 45% higher, respectively, for GAC preloaded with ozonated rather than non-ozonated water.

ACKNOWLEDGMENT

Gratefully acknowledged is the contribution of Gebereselassie Tsegaye, who conducted his M.Sc. study within this part of the research project. His good humor, and great ability and agility, were of an immense help while conducting the experiments described in this chapter.

REFERENCES

Carter, M.C., W.J.Jr. Weber and K.P. Olmstead (1992). Effect of background dissolved organic matter on TCE Adsorption by GAC. *Journal AWWA*, 8:81-91.

Carter, M.C. and W.J.Jr. Weber (1994). Modeling adsorption of TCE by activated carbon preloaded by background organic matter. *Env. Sci. Tech.*, 4:614-623.

Haist-Gulde, B. (1991). *Adsorption of trace-pollutants from surface water.* Ph.D. thesis, University Fridericana, Karlsruhe, Germany (in German).

Knappe, D.R.U., V.L. Snoeyink, Y. Matsui, M. Jose Prados and M.M. Bourbigot (1994). Determining the remaining life of a granular activated carbon (GAC) filter for pesticides. *Water Supply*, 14:1-14.

Knappe, D.R.U. (1996). *Predicting the removal of atrazine by powdered and granular activated carbon.* Ph.D. thesis, University of Illinois, Urbana, USA.

Langlais, B., D.A. Reckhow and D.R. Brink (1991). *Ozone in water treatment.* Denver, Lewis Publishers and AWWA Research Foundation.

Najm, I.N., V.L. Snoeyink and Y. Richard (1991). Effect of initial concentration of a SOC in natural water on its adsorption by activated carbon. *Journal AWWA*, 83:57-63.

Matsui, Y., J. Kamei, E. Kawase, V.L. Snoeyink and N. Tambo (1994). GAC adsorption of intermittently loaded pesticides. *Journal AWWA*, 9:91-102.

Matsui, Y. (1994). Unpublished work.

Schmidt, K.J. (1994). *Prediction of GAC column performance using bench scale techniques.* M.Sc. Thesis, University of Illinois, Urbana, USA.

Sontheimer, H., J.C. Crittenden and R.S. Summers (1988). *Activated carbon for water treatment,* AWWA - DVGW Forschungsstelle Engler Bunte Institut, Karlsruhe.

Speth T.F. and R.J. Milter (1989). Effect of preloading on the scale-up of GAC microcolumns. *Journal AWWA*, 4:141-148.

Summers, R.S., B. Haist, J. Koehler, J. Rith, G. Zimmer and H. Sontheimer (1989). The influence of background organic matter on GAC adsorption. *Journal AWWA*, 5:66-74.

Wang, G.S. and K.T. Alben (1996). Effect of preadsorbed NOM on GAC adsorption of atrazine. *Proc. First international conference on adsorption in water environment and treatment processes*, Shirahama, Japan, p. 218-225.

Weber, W.J.Jr. (1972). *Physicochemical processes for water quality control.* Wiley & Sons, New York.

Weber, W.J.Jr and B.M. van Vliet (1981). Synthetic adsorbents and activated carbons for water treatment. *Journal AWWA*, 8:420-430.

Yuasa, A. (1982). *A kinetic study of activated carbon adsorption processes.* Ph.D. thesis, Hokkaido University, Sapporo, Japan.

Zimmer, G. (1988). *Investigating the adsorption of organic trace-compounds from natural water.* Ph.D. thesis, University Fridericana, Karlsruhe, Germany (in German).

Chapter 7

Predicting Atrazine Removal by Ozone-Induced Biological Activated Carbon Filtration

ABSTRACT—Granular Activated Carbon (GAC) filters are frequently designed based on the results of the expensive and time-consuming pilot plant experiments. This makes it attractive to explore the mathematical modeling possibilities for GAC filter performance. If a target compound is not biodegraded in GAC filters, there is no essential difference between the models that can be applied to predict its removal in filters with and without enhanced bioactivity. In both cases, model parameters are determined either from the breakthrough of the target compound in pilot plant experiments or from the lab- and bench-scale experiments with preloaded GAC. Thus, in principle, currently applied adsorption models indirectly account for the biodegradation of Background Organic Matter (BOM) that takes place in biologically active filters.

Two generally available models for the adsorption in GAC filters, the simple Adams-Bohart (AB) model and the Plug Flow Homogenous Surface Diffusion (PFHSD) model, were applied to predict the breakthrough of atrazine in pilot plant GAC filters operated. The parameters of the AB model, removal capacity and removal rate, were determined from the breakthrough of atrazine observed during the initial six months of the pilot plant operation. The parameters of the PFHSD model were determined from the experiments with GAC preloaded for seven months. GAC's adsorption capacity was determined from adsorption isotherms for (pulverized) preloaded GAC, while atrazine's mass transfer coefficients were determined from the Short Fixed Bed tests with GAC particles.

The accurate prediction of the GAC filter performance proved not yet possible. The simple Adams-Bohart model allowed a close description of the initial atrazine breakthrough at the EBCT of 7 minutes but, when extrapolated, resulted in much faster breakthrough than observed during the remaining 1.5 year of the filter operation. The model was not applied for the EBCT of 20 minutes, because the effluent atrazine concentrations were too low to allow for model calibration. The PFHSD model provided a less inaccurate prediction of atrazine breakthrough at the EBCT of 7 minutes than the Adams-Bohart model, but predicted much lower breakthrough than observed at the EBCT of 20 minutes. The inaccuracy of the models applied can be explained because, due to the complexity of the processes that simultaneously take place during GAC filtration, the prediction of its performance involves many inevitable assumptions and simplifications.

7.1 INTRODUCTION

The optimum design of GAC filters, *i.e.* the one that results in the lowest cost for GAC filtration, is frequently determined from the results of pilot plant experiments. These experiments are expensive and time-consuming. This makes it attractive to explore the mathematical modeling possibilities for GAC filter performance. If mathematical models allow satisfactory predictions, they can replace pilot plant experiments, or reduce their scope and duration. In this way, both the costs and the time involved in a design of GAC filters can be reduced.

The aim of the work presented in this chapter is to determine the extent to which the breakthrough of atrazine in GAC filters with and without ozone-induced biological activity can be predicted by generally applied mathematical models. For that purpose, atrazine breakthrough observed during the two years of operation of the pilot plant GAC filters that received non-ozonated and ozonated influent spiked with 2.2 μg/l of atrazine (see Ch. 2) was compared with the prediction of the following two models:

- the Adams-Bohart model, parameters of which were determined from the initial atrazine breakthrough in the pilot plant GAC filters; and
- the Plug Flow Homogenous Surface Diffusion model (PFHSD model), parameters of which were determined from lab- and bench-scale experiments with preloaded GAC.

Note that mathematical modeling is not the only manner that can be tried for the prediction of atrazine adsorption in GAC filters. The Rapid Small Scale Column Tests (RSSCTs) are also frequently applied (Crittenden *et al.*, 1989). In these tests, GAC filters are scaled down according to the theory of adsorption, and operated with relevant water. The main advantage of these tests is that they are much cheaper and faster than the pilot plant tests. This is because the RSSCTs are conducted with much smaller volumes of water and GAC than the pilot plant tests, and in only a fraction of time required to reach the breakthrough of target compounds in the pilot plant filters. The advantage of the RSSCTs, compared to mathematical modeling, is that they directly account for the diffusion interactions between simultaneously adsorbing compounds, and for their competitive adsorption. A point that needs further research is how the RSSCTs can account for the effect of BOM preloading on the adsorption of target compound. An obvious solution is to run these tests with preloaded GAC. However, it is difficult to decide how long GAC has to be preloaded for, and to determine the effect of the pulverization (that the RSSCTs imply) of the preloaded GAC (Summers *et al.*, 1989). Another factor for which the RSSCTs do not account is the biodegradation that takes place in biologically active GAC columns.

7.2 THEORETICAL BACKGROUND

7.2.1 Mathematical modeling of the performance of GAC filters

The two main processes by which an organic compound can be removed from water passing through GAC filters are its adsorption onto GAC and its biodegradation by bacteria colonizing these filters. As discussed in Chapter 2, both are simple to describe when a single compound is present in water, and when water treated is of constant quality. The following main factors govern the removal of a single compound: (i) the adsorption capacity of GAC for that compound, (ii) the rate of the compound's external (onto GAC) and internal (into GAC) mass transfer, and (iii) the rate at which bacteria utilize the compound. Weber and Liu (1980) developed a mathematical model that includes all these factors, and showed (for sucrose and glucose) that the model results in satisfactory predictions of GAC filter performance.

If the use of such a predictive model is to be extended for a case where a target compound is present in water with other compounds, the model should predict how the presence of these compounds affects the removal of the target compound. Two elements are necessary to predict this from the individual characteristics of the compounds present in a mixture. First, all these compounds need to be identified. Secondly, the theories that describe the interactions between these compounds must be available. Unfortunately, all Background Organic Matter (BOM) compounds present in raw water are rarely identified because there are hundreds of them, some present at very low concentrations. Furthermore, the theories that would allow predictions of the three aforementioned factors (*i.e.* GAC's adsorption capacity, the compound's mass transfer rate, and the rate of compound's biodegradation) have yet to be developed. Namely, the existing theories are inadequate for the following reasons:

- while the Ideal Adsorbed Solution Theory (IAST) can be used to predict the reductions in GAC's adsorption capacity caused by competitive adsorption (see Ch. 5), it is insufficient to describe the reductions caused by the preloading of GAC (see Ch. 6);
- while external (onto GAC) mass transfer of the target compound is expected to be independent of the presence of other compounds (in dilute solutions), there is no theory for the prediction of the interactions during the internal (into GAC) mass transfer;
- the rate of compound's biodegradation in a mixture cannot be predicted from the individual biodegradation rates of all compounds present: the rate can be increased due to secondary substrate utilization, or it can be reduced due to sequential utilization (see Ch. 4).

Consequently, a model that can predict the removal of a target compound in GAC filters based on the individual characteristics of the compounds present in water treated and the changing water quality (*e.g.* temperature, pH, composition of the compounds present) is not yet available. Because of this, the adsorption capacity of GAC, the compound's mass transfer rate and the rate of its biodegradation always need to be determined from the experiments conducted with relevant water. This makes the results specific for that particular water and its quality during the experiment(s) conducted. If the intention is to account for the changing water quality, experiments need to be conducted for different sets of water quality parameters (composition of BOM mixture, temperature, pH). This increases both the costs involved in modeling and the time required to obtain model parameters from experiments.

This research focuses on the removal of a specific target compound (*i.e.* atrazine) present in pretreated Rhine River water, together with unidentified BOM compounds. In our case, the biodegradation of atrazine most likely does not contribute to its removal in GAC filters (see Ch. 4). Thus, what needs to be determined is the adsorption capacity of GAC for atrazine, the external and internal mass transfer rates of atrazine, and the effect of BOM on these factors.

Since we are modeling adsorption of atrazine in biologically active GAC filters, biodegradation of BOM plays a role in the removal of the target compound (see Ch. 4). However, when the target compound is not biodegraded in GAC filters, there is no essential difference between the mathematical models that can be applied for the prediction of its removal in filters with and without enhanced bioactivity. In both cases, model parameters are determined either from the breakthrough of the target compound in pilot plant experiments, or from the lab- and bench-scale experiments with preloaded GAC. Thus, in principle, currently applied adsorption models account indirectly for the biodegradation of BOM that takes place in biologically active filters.

Two principally different types of models can be applied to predict the performance of GAC filters. One type is the "black-box" model, which describes the removal of the target compound as a function of the removal capacity and removal rate in GAC filters. Another type of model is the one that distinguishes among the three parameters governing the adsorption in GAC filters; *i.e.* GAC's adsorption capacity as a function of the equilibrium concentration of a compound in water, the rate of the compound's external mass transfer, and the rate of the compound's internal mass transfer. Two models were chosen for use in this research project. One is a simple, "black-box" model developed in 1920 by Bohart and Adams. The other is the Plug Flow Homogenous Surface Diffusion (PFHSD) model, described by Sontheimer *et al.* (1988).

7.2.2 Adams-Bohart model

The Adams-Bohart model is the "black-box" model. However, it is simple, elegant, and was shown to allow satisfactory description of the breakthrough of both BOM (Clark, 1987) and target compounds such as phenol and methylene-blue during GAC filtration (Bendeddouche *et al.*, 1994). Note, however, that in both cases model had to be modified in order to account for less than 100% eventual breakthrough of DOC and phenol that was observed.

This model was developed by Bohart and Adams (1920) for the calculation of the service life of the activated carbon used in gas masks for the removal of chlorine from air. It assumes that the change in removal capacity over time ($\delta N/\delta t$), and the change in the concentration of the adsorbing compound over the depth of a GAC filter ($\delta C/\delta x$) are the following functions of the removal rate (k), removal capacity (N), concentration of a compound (C) and filtration velocity (v):

$$\frac{\delta N}{\delta t} = -kNC \qquad \frac{\delta C}{\delta x} = -\frac{k}{v}NC \tag{7.1}$$

The Adams-Bohart model yields the following relationship between the breakthrough of a compound (C/C_0) and the filter running time (t):

$$\frac{C}{C_0} = \left(1 + e^{kNx/v} \cdot e^{-kC_0 t} \right)^{-1} \tag{7.2}$$

Eq. 7.2 equals a simple logistic function, and results in an S-shaped curve that is symmetrical about its midpoint ($C/C_0 = 0.5$).

Parameter determination. Typically, the two model parameters (*i.e.* the removal rate and the removal capacity) are obtained by minimizing the difference between the output of the model and the breakthrough of a compound observed during a certain period of pilot plant operation. Consequently, the removal rate and the removal capacity are assumed to be constant during the whole operation. Since it is calibrated with the real breakthrough, the model indirectly accounts for the competitive adsorption and the preloading of BOM, and for the effect that biodegradation of BOM has on the two. The disadvantage of this model is that it does not give an insight into the processes taking place during adsorption in GAC filters, and that it does not allow for changing water quality.

7.2.3 Plug Flow Homogenous Surface Diffusion (PFHSD) model

The validity of the Plug Flow Homogenous Surface Diffusion (PFHSD) model for our purpose is suggested by the work of Haist-Gulde (1991), Schmidt (1994) and Knappe (1996).

They all applied the PFHSD model to describe the removal of atrazine in GAC filters receiving natural water (Rhine River, Hudson River and Seine River, respectively).

The PFHSD model includes the mathematical descriptions of the following processes (Sontheimer *et al.*, 1988): advective flow (mass transport of a compound with water being filtered); external mass transfer (by diffusion) across the hydrodynamic layer (film) around the GAC particle; local adsorption equilibrium at the particle surface; and intraparticle diffusion along the pore surface.

The PFHSD model is based on the equations for the GAC's adsorption capacity for a target compound, the compound's external (onto GAC) and internal (into GAC) mass transfer, and the mass balance equation for the packed-bed reactor, such as a GAC filter. The first three equations have been introduced in Chapter 2 (Eq. 2.1, 2.2 and 2.3, respectively). The mass balance equation is given here (Eq. 7.3). The newly introduced parameters in this equation are the porosity of GAC bed (ε) and the radius of –presumably spherical– GAC particle (R):

$$\varepsilon \frac{\partial C}{\partial t} + v \frac{\partial C}{\partial z} + \frac{3\,k_f(1-\varepsilon)}{R}(C - C_s) = 0 \tag{7.3}$$

The initial condition for equation 7.3 states that no adsorbate is originally present in the GAC column, while the boundary condition states that the adsorbate concentration at the top of the GAC adsorber equals the concentration in filter influent:

Initial condition:

$$C(z,0) = 0 \tag{7.4}$$

Boundary condition:

$$C(0,t) = C_0 \tag{7.5}$$

To obtain the effluent concentration of a target compound as a function of time, equation 7.3 has to be solved with the internal mass transfer equation 2.3 (see Ch. 2). The initial condition for this equation states that no adsorbate is present initially inside the GAC particle. The first boundary condition expresses the symmetry at the center of each GAC particle, the second equates the rate of film mass transfer at the external GAC surface to the rate of adsorbate accumulation inside the GAC particle, while the third boundary condition relates the solid- and liquid-phase concentrations at the external GAC surface:

Initial condition:

$$q(r,z,0) = 0 \qquad (7.6)$$

Boundary conditions:

$$\frac{\partial q\,(0,z,t)}{\partial r} = 0 \qquad (7.7)$$

$$D_S \rho_p \frac{\partial q\,(R,z,t)}{\partial r} = k_f[C(z,t) - C_s(z,t)] \qquad (7.8)$$

$$q_s = K C_s^n \qquad (7.9)$$

To describe the kinetics of adsorption in a GAC filter, these partial differential equations are solved numerically by methods such as orthogonal collocation (Crittenden *et al.*, 1980), or by finite differences (Yuasa, 1982).

The PFHSD model allows for the description of the competitive adsorption on GAC. Competitive adsorption is described by the Ideal Adsorbed Solution Theory (IAST) that was shown to allow, usually, an adequate description of the competitive adsorption onto virgin GAC (see Ch. 5). However, when predicting the adsorption capacity of GAC preloaded by BOM, competitive adsorption is frequently neglected. Thus, the PFHSD model is applied as if the target compound is the only compound present in water. There are two major reasons for this. First, it is likely that preloaded BOM already occupied all GAC sites it is attracted to before the target compound starts to adsorb onto GAC. Such an assumption may not always be correct, because some researchers showed that competitive adsorption does play a role on preloaded GAC (see Ch. 6). Secondly, when the competitive adsorption is neglected, the calculation is much simpler. This is because the Freundlich coefficients K and n can be assumed as constant, instead of being a function of the initial atrazine concentration.

Parameter determination. The parameters of the PFHSD model –Freundlich coefficients K and n, the film mass transfer coefficient k_f and the surface diffusion coefficient D_S – are commonly determined from bench-scale experiments with preloaded GAC. These experiments need to be done in relevant water and, if feasible, GAC should not be pulverized. GAC pulverization may enable access to GAC pores blocked by preloaded BOM and, consequently, may result in higher adsorption capacity for the target compound than the one available on intact GAC particles (see Ch. 6). Freundlich coefficients are determined from the target compound isotherm. The film mass transfer coefficient and the surface diffusion coefficient of the target compound are determined by fitting the PFHSD model to the breakthrough of the target compound in Short Fixed Bed (SFB) tests.

Because the model parameters are determined for preloaded GAC taken from either bench or pilot plant filters, the model accounts for the effect of BOM preloading, and for the effect that biodegradation of BOM has on it. Furthermore, the model can account for the effect of time and filter bed depth if GAC is sampled from the pilot plant filters at various times and from various bed depths. This, however, increases the number of the experiments required to determine model parameters. A disadvantage of this model is that it does not account for the changing water quality (*e.g.* composition of BOM mixture, temperature and pH).

7.3 MATERIALS AND METHODS

7.3.1 Adams-Bohart model

Parameter determination. Adsorption capacity and adsorption rate parameters of the Adams-Bohart model were determined by fitting the linear form of this model (Eq. 7.10) to the breakthrough of atrazine observed at a bed depth of 0.35 m (EBCT 7 min) during the initial six months of the pilot plant operation. The same could not be done for a bed depth of 1.1 m (EBCT 20 min), because atrazine concentrations in the effluent were mostly below the detection limit of the analytical method applied (0.03 μg/l).

$$ln\left(\frac{C_0}{C} - 1\right) = \frac{kNx}{v} - kC_0t \qquad (7.10)$$

Table 7.1 Operational parameters for pilot plant GAC filters.

Parameter	non-ozonated influent	ozonated influent
C_0 (μg atrazine/l)	2.22 (±0.24)	2.24 (±0.22)
approach velocity (cm/min)	5.45	5.45
equivalent sphere diameter (cm)	0.104	0.104
apparent particle density (g/cm^3)	0.380	0.380
bulk bed density (g/cm^3)	0.334	0.334
total GAC filter running time (days)	800	800
total GAC bed depth (cm)	35 (110)	35 (110)
GAC filter diameter (cm)	26.5	26.5

Prediction. Model parameters obtained in the above manner, and the operational parameters of the pilot plant GAC filters shown in Table 7.1, were used to extrapolate the Adams-Bohart model for the description of the breakthrough of atrazine (at a bed depth of 0.35 m) during the remaining 1.5 year of the pilot plant operation.

7.3.2 Homogenous Surface Diffusion (PFHSD) model

Parameter determination. Parameters of the PFHSD model were determined for GAC that was preloaded for 7 months with either non-ozonated or ozonated pretreated Rhine River water. GAC was preloaded in the pilot plant GAC filters described in Chapter 6. Freundlich coefficients K and n have been determined from atrazine isotherms (C_0 of 3 μg/l) for pulverized GAC. Atrazine mass transfer coefficients k_f and D_s have been determined by fitting the pseudo single-solute PFHSD model to the breakthrough of atrazine observed in the Short Fixed Bed tests conducted with intact GAC particles (Table 7.2). Both the adsorption isotherm tests and the SFB tests were done with the same water used for GAC preloading, *i.e.* either ozonated or non-ozonated pretreated Rhine River water, and following the same procedure as the one described in Chapter 6.

Table 7.2 Operational parameters for Short Fixed Bed tests with preloaded GAC.

Water used for GAC preloading and SFB test	C_0 (μg atr./l)	mass of GAC (g)	GAC bed depth (cm)	Q (l/h)
non-ozonated water	3.0	32.1	4.9	6.4
ozonated water	2.7	32.1	4.9	6.4

Prediction. Model parameters obtained in the above manner, and the operational parameters of the pilot plant GAC filters shown in Table 7.1, were used to predict the breakthrough of atrazine in the pilot plant GAC filters by the pseudo single-solute PFHSD model. Model equations were solved using the program written by Yuasa (1982). This program uses a moving-grid finite difference method to solve the PFHSD model equations numerically.

Determination of Freundlich coefficient K required. The PFHSD model was solved for various values of Freundlich coefficient K, while keeping other model parameters constant. The model was run with the Freundlich n of 0.41, which was obtained for atrazine isotherm (C_0 of 3 μg/l) with virgin GAC and non-ozonated pretreated Rhine River water (Table 5.2), and external and internal mass transfer rate coefficients of $15 \cdot 10^{-4}$ and $9.6 \cdot 10^{-13}$ obtained from the SFB tests for virgin GAC and demineralized water (Table 6.3). This was done to determine which percentage of the initial Freundlich coefficient K (for virgin GAC) represents roughly the average coefficient K available in the pilot plant filters during the two years of their operation.

7.4 RESULTS AND DISCUSSION

7.4.1 Adams-Bohart model

The Adams-Bohart model allowed for an accurate description of the breakthrough of atrazine observed at the EBCT of 7 minutes during the initial six months of pilot plant operation (Fig. 7.1). This model provided accurate descriptions for both GAC filters, the one that received non-ozonated and the one that received ozonated water. The removal capacity and removal rate obtained for each filter are given in Table 7.3.

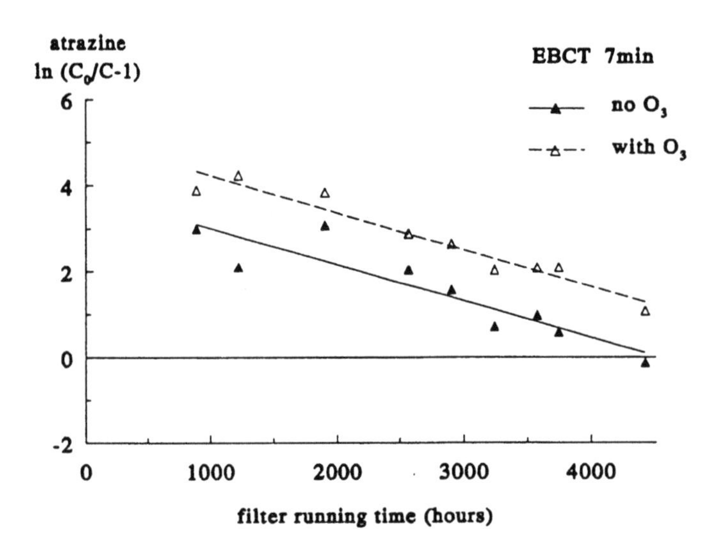

Figure 7.1 Fit of the Adams-Bohart model to the breakthrough of atrazine observed at 7 minutes EBCT during the initial six months of GAC filters operation (points).

Table 7.3 Coefficients of the Adams-Bohart model determined by fitting the model to atrazine breakthrough (bed depth 0.35 m) during the initial six months of filter operation (C_0 of 2 μg/l).

Adams-Bohart coefficients	kNx/v (-)	kC_0 (h^{-1})	k (m^3/g h)	N (g/m^3)
non-ozonated water	3.861	$0.852 \cdot 10^{-3}$	0.426	85.47
ozonated water	5.099	$0.863 \cdot 10^{-3}$	0.432	111.29

Although the Adams-Bohart model described the initial breakthrough quite well, there was a significant difference between the atrazine breakthrough observed during the two years of pilot plant operation and the one obtained by extrapolating the Adams-Bohart model (Fig. 7.2).

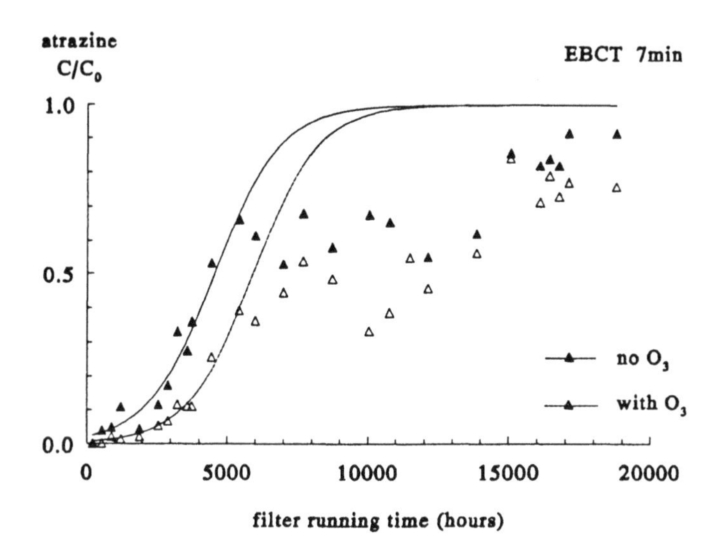

Figure 7.2 Breakthrough of atrazine at 7 minutes EBCT, as predicted by the Adams-Bohart model.

The model predicted a lower removal capacity of GAC for atrazine than actually available (more atrazine was retained in the filter than predicted by the model), and a faster removal rate than the actual one (steeper breakthrough curve predicted than observed). Such a discrepancy can, possibly, be explained by changed quality of the raw water after the initial six months during which the model was calibrated. The improvement in the quality of pretreated Rhine River water may be expected, considering the increasing efforts to reduce the micropollutant-loadings it receives.

7.4.2 PFHSD model

The parameters of this model were determined for GAC preloaded for seven months with either ozonated or non-ozonated pretreated Rhine River water. In Table 7.4 given are the adsorption capacity parameters (K, n) and the mass transfer parameters (k_f, D_S) that were obtained from adsorption isotherms (Fig. 7.3) and the SFB tests (Fig. 7.4), respectively.

The breakthroughs of atrazine at the EBCT of 7 minutes and 20 minutes, obtained when these parameters were incorporated into the PFHSD model, are shown in Figure 7.5 and Figure 7.6, respectively.

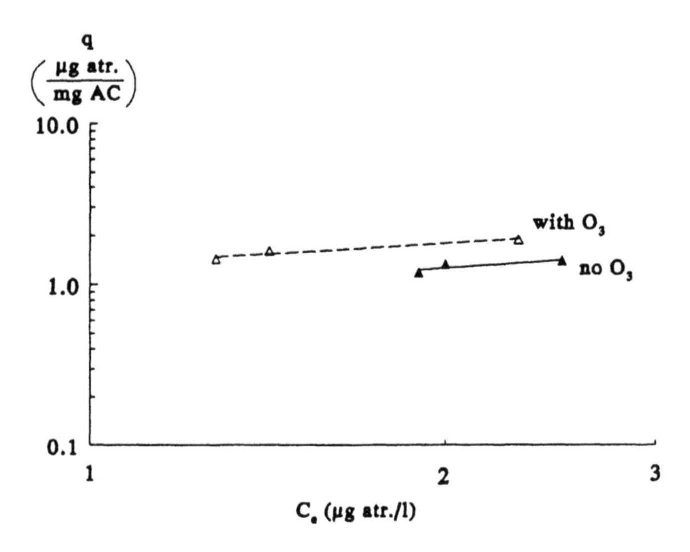

Figure 7.3 Atrazine isotherms for GAC preloaded for seven months with either non-ozonated or ozonated pretreated Rhine River water.

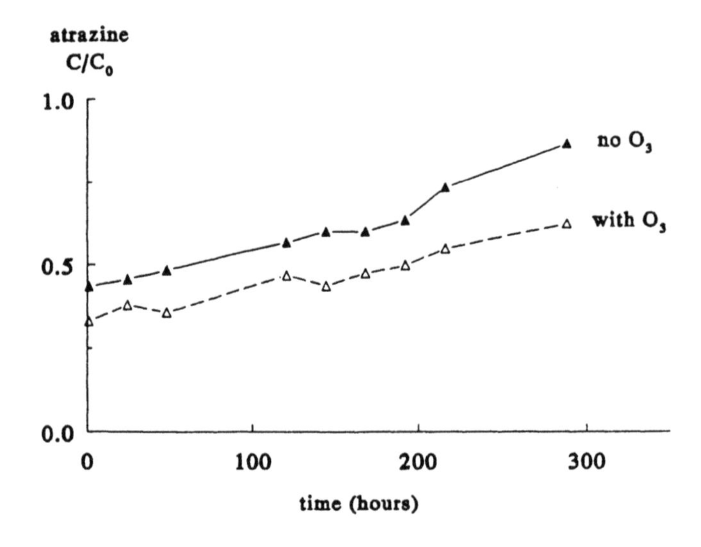

Figure 7.4 Atrazine breakthrough in the SFB tests for GAC preloaded for seven months with either non-ozonated or ozonated pretreated Rhine River water.

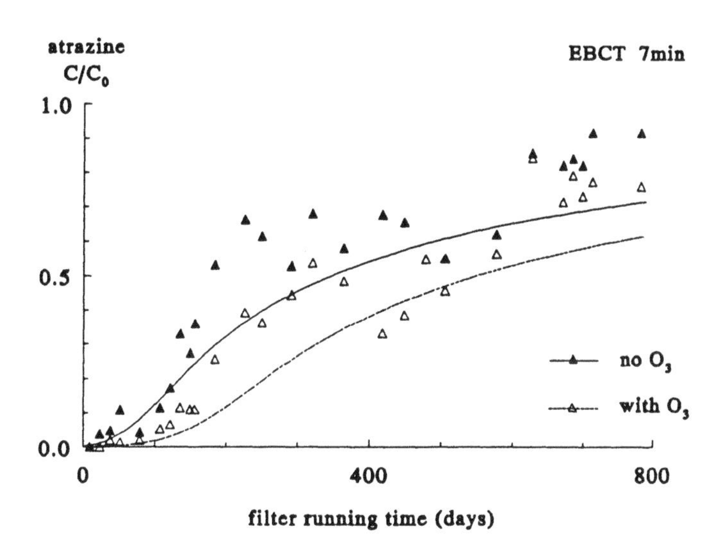

Figure 7.5 Atrazine breakthrough at 7 minutes EBCT, as predicted by the PFHSD model using adsorption capacity and mass transfer parameters determined for GAC preloaded for seven months.

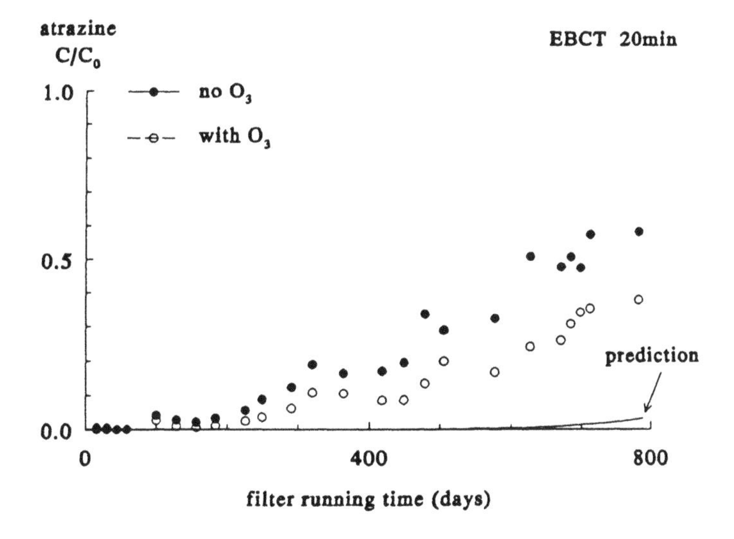

Figure 7.6 Atrazine breakthrough at 20 minutes EBCT, as predicted by the PFHSD model using adsorption capacity and mass transfer parameters determined for GAC preloaded for seven months.

Table 7.4 Coefficients of the PFHSD model determined from adsorption isotherms and SFB tests with GAC preloaded for 7 months with either ozonated or non-ozonated pretreated Rhine River water.

Parameter	GAC preloaded with non-ozonated water	GAC preloaded with ozonated water
K (mg/g)(μg/l)$^{-n}$	0.89	1.34
n (-)	0.51	0.42
k_f (cm/s)	$3.12 \cdot 10^{-4}$	$3.97 \cdot 10^{-4}$
D_S (cm^2/s)	$16.28 \cdot 10^{-13}$	$14.24 \cdot 10^{-13}$

For the bed depth of 0.35 m (EBCT 7 min), the PFHSD model resulted in less inaccurate prediction than the simple Adams-Bohart model. However, for the bed depth of 1.1 m (EBCT 20 min), the PFHSD model predicted the breakthrough of only a few percent at the end of the two years of filter operation. In contrast, the actual atrazine breakthrough amounted to 40% and 60% in the filter receiving ozonated and non-ozonated influent, respectively.

There are various reasons that can explain this inaccurate prediction of the PFHSD model. First, the model does not account for the change in water temperature during the pilot plant operation. Model parameters were namely determined from a single experiment conducted at 20°C. This is an important limitation because an increase in water temperature may cause desorption of previously adsorbed compounds and, consequently, may reduce the internal mass transfer of atrazine more than predicted.

Secondly, it is possible that atrazine and BOM compete for the adsorption sites available on preloaded GAC. Based on the findings of Knappe *et al.* (1994), who worked with atrazine and pretreated Seine River water, we assumed that there is no competitive adsorption on GAC preloaded with pretreated Rhine River water, either. Furthermore, the software that was available for our research does not allow for this competitive adsorption. However, if BOM and atrazine compete for the adsorption on preloaded GAC, the adsorption capacity of GAC varies with the initial atrazine concentration. Thus, deeper in a filter bed where atrazine concentration is lower than in filter influent, both the adsorption capacity of GAC and the mass transfer rate of atrazine are strongly reduced. This results in faster atrazine breakthrough than the one predicted assuming that there is no competitive adsorption.

Finally, although the average influent DOC concentration in both experiments was 2.1 mg/l, it is possible that the composition of BOM in pretreated Rhine River water was not the same during the pilot plant operation (1993-1995) and during the preloading of GAC (1996). If there was a difference, the reduction in GAC's capacity for atrazine and in atrazine's mass transfer rate may have been more pronounced during the pilot plant operation than during GAC preloading. In addition, seven months of preloading may have been too short a period to reduce the adsorption capacity and mass transfer coefficients to the extent required to approximate the breakthrough of atrazine during the two years of pilot plant operation. Also, the adsorption capacity was determined for pulverized GAC and, therefore, was probably too high. Though this may have been compensated by a reduction in mass transfer coefficients, which were deduced from atrazine breakthrough in the SFB tests assuming such overestimated adsorption capacity

Adsorption capacity required. Figure 7.7 shows that the breakthrough of atrazine in GAC filters receiving non-ozonated and ozonated pretreated Rhine River water can be roughly approximated when applying the Freundlich *K* of 40% and 50%, respectively, of the one obtained from atrazine isotherm (C_0 of 3 μg/l) with virgin GAC and non-ozonated Rhine River water (Figure 5.12). Such a reduction of the initial coefficient *K* for atrazine (the one obtained for virgin GAC) is not unrealistic, since it is comparable to the reductions observed elsewhere (Haist-Gulde, 1991; Knappe *et* al., 1994; Wang and Alben, 1996).

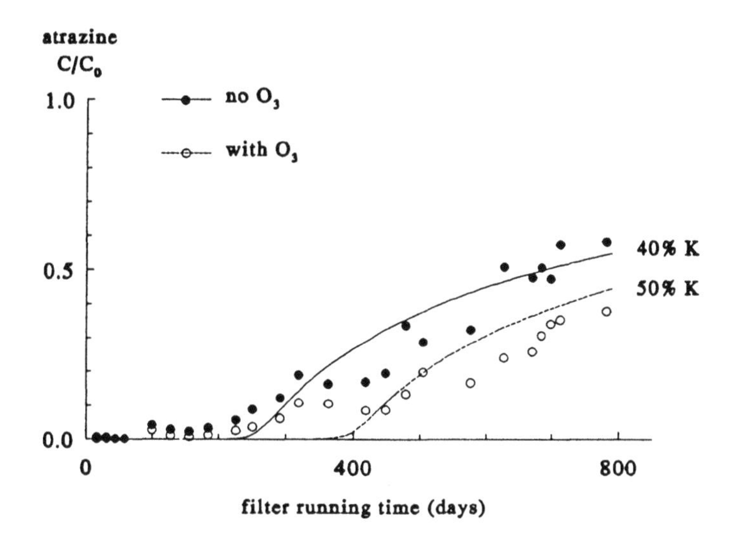

Figure 7.7 Atrazine breakthrough (EBCT 20 min) obtained by the PFHSD model when using 40% and 50% of the initial *K* for atrazine isotherm (C_0 of 3μg/l) in non-ozonated pretreated Rhine River water.

CONCLUSIONS

The simple Adams-Bohart model allows for an accurate description of the breakthrough of atrazine (EBCT 7 min) observed during the initial stages of pilot plant operation. However, the model calibrated in this way cannot be extrapolated to describe atrazine breakthrough after that. In this research, the predicted breakthrough was much faster than the one observed. Such inaccurate extrapolation can possibly be explained by a change in water quality after the initial six months during which the model was calibrated. The model was not applied for atrazine breakthrough at the EBCT of 20 minutes. Namely, during the initial six months of the pilot plant operation, atrazine concentrations in filter effluent were too low (they were mostly below the detection limit) to allow for model calibration.

The PFHSD model run with the adsorption capacity and mass transfer coefficients determined for GAC preloaded for seven months does not allow accurate prediction of atrazine breakthrough. At the EBCT of 7 minutes the prediction of this model was less inaccurate than the extrapolation made by the Adams-Bohart model. However, for the EBCT of 20 minutes, the PFHSD model predicted only a few percent (< 5%) atrazine breakthrough after two years of filter operation, while 60% and 40% breakthroughs were observed in the filters that received non-ozonated and ozonated influent, respectively.

It is difficult to attribute the inaccurate prediction of the PFHSD model to a single cause. However, it can be explained in many ways. For example, the model neglects reductions in the internal mass transfer rate of atrazine caused by desorption of other compounds in GAC filters (*e.g.* brought about by increased water temperature or a change in water quality), while the temperature of Rhine River water during the course of this research ranged from 2°C to 21°C. In addition, the model works with Freundlich coefficients K and n that are constant and, thus, independent of the initial atrazine concentration. This cannot be assumed valid if BOM and atrazine compete for the adsorption on preloaded GAC. It is also possible that the quality of Rhine River water was not the same during the pilot plant operation (1993-1995) and during the preloading of the GAC used to determine model parameters (1996). Finally, seven months of preloading may have been too short a period to reduce the adsorption capacity and mass transfer parameters to the extent required to approximate the breakthrough of atrazine during the two years of pilot plant operation.

An accurate prediction of the GAC filter performance is not yet possible. This can be expected considering that, due to the complexity of the processes that simultaneously take place during GAC filtration, such prediction involves many inevitable assumptions and simplifications.

SYMBOLS

C_0 = influent concentration of a compound ($kg \cdot m^{-3}$)
C = liquid phase concentration of a compound ($kg \cdot m^{-3}$)
C_s = liquid phase concentration at the external surface of GAC particle ($kg \cdot m^{-3}$)

k = removal rate ($m^3 \cdot kg^{-1} \cdot s^{-1}$)
N = removal capacity ($kg \cdot m^{-3}$)
q_s = solid-phase concentration at the external surface of GAC (-)
x = GAC bed depth (m)
v = filtration velocity assuming an empty filter ($m \cdot s^{-1}$)
t = filter running time (s)
z = axial position within the GAC filter (m)
ε = bed porosity (-)
k_f = film mass transfer coefficient ($m \cdot s^{-1}$)
R = radius of GAC particle (m)

REFERENCES

Bendeddouche, N., H. Benhocine, F. Kaoua and A. Gaid (1994). Modeling removal of organic compounds by activated carbon using a logistic function. *JWSRT-Aqua, 43,* 17-21 (in French).

Bohart, G.S. and E.Q. Adams (1920). Some aspects of the behavior of charcoal with respect to chlorine. *Journal Am.Chem.Soc.*, 42:523-544.

Clark, R.M. (1987). Modeling TOC removal by GAC: the general logistic function. *Journal AWWA,* 1:33-37.

Crittenden, J.C., B.W.C. Wong, W.E. Thacker, V.L. Snoeyink and R.L. Hinrichs (1980). Mathematical model of sequential loading in fixed-bed adsorbers. *Journal WPCF,* 11:2780-2795.

Crittenden, J.C., P.S. Reddy, D.W. Hand and H. Arora (1989). *Prediction of GAC performance using rapid small-scale column tests.* AWWA Research foundation, Denver.

Haist-Gulde, B. (1991). *Adsorption of trace-pollutants from surface water.* Ph.D. thesis, University Fridericana, Karlsruhe, Germany.

Knappe, D.R.U., V.L. Snoeyink, Y. Matsui, M. Jose Prados and M.M. Bourbigot (1994). Determining the remaining life of a granular activated carbon (GAC) filter for pesticides. *Water Supply,* 14:1-14.

Knappe, D.R.U. (1996). *Predicting the removal of atrazine by powdered and granular activated carbon.* Ph.D. thesis, University of Illinois, Urbana, USA.

Schmidt, K.J. (1994). *Prediction of GAC column performance using bench scale techniques.* M.Sc. thesis, University of Illinois, Urbana, USA.

Sontheimer, H., J.C. Crittenden and R.S. Summers (1988). *Activated carbon for water treatment,* AWWA - DVGW Forschungsstelle Engler Bunte Institut, Karlsruhe.

Summers, R.S., B. Haist, J. Koehler, J. Ritz, G. Zimmer and H. Sontheimer (1989). The influence of background organic matter on GAC adsorption. *Journal AWWA,* 5:66-74.

Wang, G.S. and K.T. Alben (1996). Effect of preadsorbed NOM on GAC adsorption of atrazine. *Proc. First international conference on adsorption in water environment and treatment processes,* Shirahama, Japan, p. 218-225.

Weber, W.J. Jr. and K.T. Liu (1980). Determination of mass transport parameters for fixed-bed adsorbers. *Chem. Eng. Comm.,* 6:264-270.

Yuasa, A. (1982). *A kinetic study of activated carbon adsorption processes.* Ph.D. thesis, Hokkaido University, Sapporo, Japan.

Chapter 8

Summary

8.1 INTRODUCTION

Since the seventies, new water treatment processes have been introduced in the production of drinking water from surface water. Their major aim was to ensure the adequate disinfection of this water and/or the efficient removal of pesticides and other organic micropollutants from it.

Amsterdam Water Supply (AWS) recently developed two new integral concepts for the treatment of surface water (*i.e.* Rhine River). In contrast to water treatment schemes typically applied in The Netherlands, these two new concepts do not use a reservoir or artificial recharge to provide for the periods when the Rhine River is heavily polluted, or to improve the quality of water. Thus, these two concepts have to offer ultimate barriers against micropollutants and microorganisms present in Rhine River water, both under normal and accidental conditions. In addition, they have to reduce the high salinity, in particular the high chloride concentration, and the hardness of the raw water. These two new concepts include conventional pretreatment by coagulation, sedimentation and rapid sand filtration, followed either by slow sand filtration and reverse osmosis or by ozonation, Granular Activated Carbon (GAC) filtration, slow sand filtration and reverse osmosis. AWS tested and evaluated the performances of these two concepts for the future capacity extension at the AWS' Leiduin plant with the additional $13 \cdot 10^6$ m^3 per year. Particular attention has been paid to the following aspects: (i) removal of pesticides, metabolites and other organic micropollutants by combined ozonation and GAC filtration, and by reverse osmosis, (ii) disinfection by ozonation and by reverse osmosis, and (iii) control of the fouling and scaling of reverse osmosis membranes.

The mechanisms that play a role in these processes were studied in the context of a research project conducted jointly by IHE, NORIT NV, Kiwa NV and AWS. The research presented in this thesis was conducted within the framework of this project. Its focus is on Biological Activated Carbon (BAC) filtration, which is a combination of ozonation and Granular Activated Carbon filtration. The general goal of this research is to identify and understand mechanisms that underlie the expected beneficial effect of ozonation on the removal of pesticides and other micropollutants by GAC filtration. This understanding allows one to judge whether a combination of ozonation and GAC filtration provides a sound barrier against these compounds and, in addition, allows optimization of the underlying mechanisms.

Detailed investigations for all pesticides and other organic micropollutants present in Rhine River water were not possible within the scope of this research project. Such investigations would be too costly and, in addition, the appropriate analytical methods were not available for many of these compounds. Thus, the research was done for one model compound - atrazine. Pesticide atrazine was chosen as a model compound because, at the start of this research in 1992, its removal was particularly relevant for AWS; it was detected in pretreated Rhine River water at concentrations higher than the European Union standard of 0.1 μg/l. Moreover, atrazine was judged to be reasonable resistant to biodegradation. Its enhanced biodegradation in GAC filters receiving ozonated rather than non-ozonated influent would therefore, if verified, indicate enhanced biodegradation of other pesticides as well. Last but not the least, the analytical method needed to measure atrazine concentrations below 0.1 μg/l was available.

8.2 PROCESS ANALYSIS AND RESEARCH SCOPE

Before the scope of this research was defined, the processes anticipated as playing a role in the removal of pesticides by BAC filtration were analyzed (via desk study) and preliminary experiments were conducted (see Ch. 2). One their objective was to relate the oxidation of atrazine and the formation of its oxidation byproducts to the ozone dose applied, and to indicate the efficiency at which these byproducts are removed in GAC filters. The main purpose, however, was to verify the expectation that the removal of atrazine in GAC filters is improved due to ozone-induced oxidation of Background Organic Matter (BOM) from pretreated Rhine River water. BOM refers to the organic matter in the influent of GAC filters other than the target compounds that need to be removed. BOM is mostly of natural origin, *e.g.* compounds such as humic substances, but it also includes –especially BOM present in Rhine River water– many compounds of anthropogenic origin. This process analysis and the experiments led to following conclusions:

- Ozonation can be expected to improve the removal of pesticides by GAC filtration due to two effects. One is the well-known effect of oxidation of pesticides. Another is the effect of oxidation of Background Organic Matter (BOM) present in filter influent. Namely, an important part of BOM compounds will be partially oxidized because of ozonation. This oxidation increases the biodegradability, and reduces the adsorbability and molecular mass of these compounds. Consequently, enhanced biodegradation and reduced adsorption of BOM are expected in filters receiving ozonated rather than non-ozonated influent. Both biodegradation and adsorption of pesticides are expected to be improved as a result.

- Ozonation of pretreated Rhine River water results in limited oxidation of atrazine: about 25%, 45% and 65% of atrazine are oxidized when applying an ozone dose of 0.5 mg/l, 1 mg/l and 1.5 mg /l, respectively. As frequently shown in other studies, the resulting decreased atrazine concentration in the influent of GAC filters delays its breakthrough. Ozonation also results in the formation of desethylatrazine and desisopropylatrazine. These two byproducts of atrazine oxidation are formed at concentrations up to 20% and 5% of the original atrazine concentration, respectively. Most likely, other oxidation byproducts that were not monitored (*e.g.* hydroxyatrazine, desisopropylatrazine amide, etc.) are formed also. The two byproducts monitored are more biodegradable but, as shown in this study, less adsorbable than atrazine. Because of these characteristics, it is difficult to predict whether they are removed by GAC filtration better than atrazine itself.

- Removal of atrazine by GAC filtration is significantly improved due to ozone-induced oxidation of BOM from pretreated Rhine River water. This was concluded because improved atrazine removal was observed in the filter that received ozonated rather than non-ozonated influent, while both influents had the same concentration of atrazine (*i.e.* atrazine was spiked after complete depletion of ozone). Based on the available analytical capacity, removal of atrazine was monitored for two empty-bed-contact-times (7 and 20 minutes). Due to this effect of ozonation, GAC in filters receiving ozonated influent can be regenerated less frequently. This results in important savings. For example, for the treatment capacity of $13 \cdot 10^6$ m^3 per year and for the conditions tested (atrazine concentration 2.2 µg/l, ozone dose 0.8 mg/l, EBCT 20 minutes), ozonation reduces the annual cost of GAC regeneration for DFL 60,000 and DFL 130,000 when 30% and 10% breakthrough of atrazine are used as the regeneration criterion, respectively.

Once this expectation was verified, the scope of further research was defined, with the following specific objectives:

- to assess the range of ozone doses that may be applied regarding the role of ozonation;
- to determine the role of biodegradation in the removal of atrazine in GAC filters;
- to quantify the extent to which ozonation reduces the competitive adsorption of BOM;
- to quantify the extent to which ozonation reduces the preloading of BOM; and
- to determine the extent to which the breakthrough of atrazine in GAC filters with and without ozone-induced bioactivity can be predicted by commonly applied models.

8.3 DISINFECTION AND FORMATION OF BIODEGRADABLE ORGANIC MATTER AND BROMATE

Several pilot plant experiments were done to assess the range of ozone doses that may be applied when the role of ozonation is to provide an essential contribution to the disinfection of water and to promote biodegradation in GAC filters, while avoiding an unacceptable formation of bromate (see Ch. 3). Such an assessment also defines the range of ozone doses that are relevant for this research, and for which experiments need to be done. Keeping in mind that the results were obtained for pretreated Rhine River water (DOC≈2 mg/l, Br^- from 150 µg/l to 200 µg/l, pH from 7.5 to 8.0), and that they are not necessarily applicable for waters of different quality, the following was concluded:

- As shown by all the experiments conducted, ozone doses lower than 0.8 mg/l result in less than 10 µg/l of bromate, which is the standard proposed for the European Union. Ozone doses up to 3 mg/l can be applied when taking into account that about 96% of bromate formed is removed by reverse osmosis, planned as the final step for the extension of the treatment capacity at the AWS' Leiduin plant. Due to the significant contribution of OH radicals to the formation of bromate, the model that accounts just for the reactions of molecular ozone does not allow accurate predictions for the conditions at AWS.

- The concentration of biodegradable organic matter, measured as the Assimilable Organic Carbon (AOC) concentration, is increased by ozonation. An ozone dose of 0.75 mg/l already increases the AOC concentration for more than 50% of the maximum 170 µg Ac-C/l formed for ozone doses up to 3 mg/l.

- When complete mixing is assumed in ozone contact tanks, an ozone dose of 1.5 mg/l may be expected to result in the Ct value sufficient for, at least, 2-log inactivation of viruses and 1-log inactivation of *Giardia* cysts. An ozone dose of 3 mg/l may be expected to result in at least 14-log and 7-log inactivation of viruses and *Giardia* cysts, respectively. Disinfection in the full-scale plant will be higher than this, depending on the degree to which the short-circuiting in ozone contact tanks can be limited.

After the range of applicable ozone doses was defined, the experiments were done to identify processes that underlie improved removal of atrazine observed in filters that received ozonated rather than non-ozonated influent. It was also aimed to quantify the contribution of each of these processes.

8.4 BIODEGRADATION AND ATRAZINE REMOVAL BY GAC FILTRATION

The specific objective of the experiments done in this context was to verify the expected enhanced biodegradation of BOM in GAC filters receiving ozonated influent, and to find out whether it improves biodegradation and/or adsorption of atrazine in these filters (see Ch. 4). The pilot and bench scale experiments, conducted with both pretreated Rhine River water and synthetic water, resulted in the following conclusions:

- Improved removal of BOM observed in filters that received ozonated influent can be attributed to the enhanced biodegradation of BOM in these filters. This can be concluded because ozonated, rather than non-ozonated BOM, was also better removed in Non-Activated Carbon (NAC) filters, in which the removal of BOM is via biodegradation only.

- It could not be demonstrated that biodegradation of atrazine accounts for its improved removal in GAC filters that received ozonated rather than non-ozonated influent. Namely, no indication of atrazine biodegradation in these GAC filters was found in either of the experiments conducted:
  - no metabolites of atrazine biodegradation were detected in the effluent of atrazine-spiked GAC filters;
  - atrazine was not removed in NAC filters;
  - atrazine was not removed in the liquid media inoculated with the bacteria taken from atrazine-spiked GAC filters;
  - after three years of pilot plant operation, more atrazine was desorbed from GAC taken from the filter that received ozonated rather than non-ozonated influent.

- These results, however, do not entirely exclude atrazine biodegradation in GAC filters. Atrazine biodegradation was demonstrated during this research for lab-scale columns filled with glass beads while Huang and Banks (1996) demonstrated it for lab-scale columns filled with GAC. Thus, bacteria can be expected to biodegrade atrazine in GAC filters where the same conditions are created. The conditions in the pilot-plant GAC filters operated within this research, for which biodegradation of atrazine could not be demonstrated, were however quite different than those in the aforementioned lab-scale experiments. Namely, atrazine concentration in the pilot-plant experiment was 2 μg/l while atrazine concentration in the lab-scale experiments was 15 mg/l (glass beads experiment) and 50 μg/l (GAC experiment). Furthermore, water temperature in the pilot-plant experiment varied between 2°C and 21°C, while water temperature in the lab-scale experiment with glass beads was constant at a level of 30°C and in the experiment with GAC was (presumably) equal to room temperature. Finally, in contrast to the pilot GAC filters, the lab-scale columns with glass beads were inoculated with atrazine-degrading bacteria and were operated with the addition of basic salts and micronutrients to their influent, while the lab-scale columns with GAC were inoculated with bacteria taken from water and sediment of water reservoir with long history of exposure to atrazine (to concentrations up to 50 μg/l) and were operated with effluent-recycling.

- The enhanced biodegradation of BOM in filters receiving ozonated influent improves adsorption of atrazine in GAC filters. This can be concluded because atrazine was found better adsorbed onto GAC preloaded with ozonated water that passed through Non-Activated Carbon (NAC) filters, than onto GAC preloaded directly with ozonated water. Because of negligible adsorption of BOM in NAC filters, only biodegradation of BOM in NAC filters could account for the improved adsorption of atrazine that was observed. However, reduced adsorbability and molecular mass of ozonated BOM compounds can also contribute to the improved adsorption of atrazine in GAC filters.

8.5 COMPETITIVE ADSORPTION OF BOM

The experiments done within this context aimed to quantify the extent to which ozonation reduces the competitive adsorption of BOM and atrazine, due to reduced adsorbability of oxidized BOM compounds from pretreated Rhine River water (see Ch. 5). The effect of enhanced biodegradation of ozonated BOM, which is also expected to reduce the competitive adsorption of BOM, was not investigated. The following was concluded:

- Ozonation significantly increases the adsorption capacity of GAC for atrazine, due to the reduced adsorbability of oxidized Background Organic Matter (BOM) compounds. The same GAC's adsorption capacity for atrazine is obtained for ozone doses between 1 mg/l and 4 mg/l. This is remarkable because DOC isotherms showed that an increase in ozone dose further reduces the adsorbability of BOM. It is possible that after a certain ozone dose (*e.g.* 1.1 mg/l), ozonation further reduces only the adsorbability of BOM compounds that do not compete with atrazine. At the equilibrium atrazine concentration of 0.1 μg/l, the adsorption capacity in ozonated water is roughly doubled compared with the capacity in non-ozonated water. This was concluded from the isotherms determined for the initial atrazine concentration of 3 μg/l, which was the lowest atrazine concentration for which isotherms could be determined from the experiments.

- The Fictive Component Adsorption Analysis (FCAA) method and the Equivalent Background Compound (EBC) method, combined with the Ideal Adsorbed Solution Theory (IAST), provide reasonable estimates for the competitive adsorption of atrazine and BOM from non-ozonated pretreated Rhine River water. This is in contrast to Freundlich-IAST method. Such conclusion was drawn from the experiments done at the initial atrazine concentrations from 30 μg/l to 300 μg/l. Therefore, the FCAA-IAST method and the EBC-IAST method can be expected to allow predictions of GAC's adsorption capacity at initial atrazine concentrations that are too low to do the experiments.

- The FCAA-IAST and the EBC-IAST methods predict that the equilibrium (C_e = 0.1 μg/l) adsorption capacity for atrazine in non-ozonated pretreated Rhine River water at the initial atrazine concentration of C_0 = 0.3 μg/l is about 50% of the one obtained from the experimental isotherm determined at C_0 = 3 μg/l. Furthermore, this capacity is only a fraction (from 2% to 5%) of the one obtained from the common isotherm tests done at C_0 between 30 μg/l and 300 μg/l. This is important, because the concentration of 0.3 μg/l was the highest atrazine concentration measured in pretreated Rhine River water in the period 1991-1995.

- The EBC-IAST method –based on the analysis of atrazine isotherms– can be preferred for the prediction of the adsorption capacity for atrazine in ozonated water. This method predicts that the equilibrium (C_e = 0.1 μg/l) adsorption capacity for atrazine at C_0 = 0.3 μg/l is about two times higher in ozonated rather than non-ozonated pretreated Rhine River water.

8.6 BOM PRELOADING

The purpose of the experiments done within this context was to verify that the preloading of BOM present in pretreated Rhine River water speeds up the breakthrough of atrazine from GAC filters, and that ozonation reduces this adverse impact of BOM preloading. In addition, it was aimed to quantify the effect of BOM preloading on the parameters that govern the adsorption of atrazine in GAC filter (*i.e.* GAC's adsorption capacity for atrazine, and the coefficients of both external and internal mass transfer rate of atrazine), and to quantify the effect of ozonation on these parameters (see Ch. 6). The following was concluded:

- Preloading of BOM from pretreated Rhine River water speeds up the breakthrough of atrazine in GAC filters. This finding is corroborated by the results of other studies, conducted for many adsorbates and raw water sources. The effect is more pronounced for longer preloading times. Ozonation reduces this adverse effect of preloading: atrazine breakthrough for GAC preloaded with ozonated rather than non-ozonated water was found to be 17% and 26% lower after two weeks and after eight weeks of preloading, respectively.

- The faster breakthrough of atrazine in filters preloaded with BOM is due to the lower adsorption capacity of GAC for atrazine, the lower external mass transfer rate of atrazine and the lower internal mass transfer rate of atrazine at preloaded GAC than at virgin GAC. Ozonation reduces this adverse effect of preloading. GAC preloaded for two weeks with ozonated rather than non-ozonated water was found to have 19% higher adsorption capacity for atrazine (measured as Freundlich coefficient K), and 20% higher external mass transfer coefficient for atrazine. After eight weeks of preloading, GAC's adsorption capacity and the external and internal mass transfer coefficients of atrazine were 32%, 60% and 45% higher, respectively, for GAC preloaded with ozonated rather than non-ozonated water.

8.7 PREDICTING ATRAZINE REMOVAL BY OZONE-INDUCED BAC FILTRATION

The last objective of this research was to determine the extent to which the breakthrough of atrazine observed in GAC filters with and without ozone-induced bioactivity can be predicted by generally applied mathematical models (see Ch. 7). Assuming that atrazine was not biodegraded in GAC filters these models can be applied since, if a target compound is not biodegraded in GAC filters, there is no essential difference between the mathematical models that can be applied for the prediction of its removal in filters with and without enhanced bioactivity. In both cases, model parameters are determined from the breakthrough of the target compound in pilot plant experiments or from the lab- and bench-scale experiments with preloaded GAC. Thus, in principle, currently applied adsorption models indirectly account for the biodegradation of BOM that takes place in biologically active filters.

The following was concluded:

- The simple Adams-Bohart model allows for an accurate description of the breakthrough of atrazine (EBCT 7 min) observed during the initial stages of pilot plant operation. However, the model calibrated in this way cannot be extrapolated to describe atrazine breakthrough after that. In this research, the predicted breakthrough was much faster than the one observed. Such inaccurate extrapolation can possibly be explained by a change in water quality and/or temperature after the initial six months during which the model was calibrated. The model was not applied for atrazine breakthrough at the EBCT of 20 minutes. Namely, during the initial six months of the pilot plant operation, atrazine concentrations in filter effluent were too low (they were mostly below the detection limit) to allow for model calibration.

- The PFHSD model run with the adsorption capacity and mass transfer coefficients determined for GAC preloaded for seven months does not allow accurate prediction of atrazine breakthrough. The prediction of this model was less inaccurate at the EBCT of 7 minutes than the extrapolation made by the Adams-Bohart model. However, for the EBCT of 20 minutes, the PFHSD model predicted only a few percent (< 5%) atrazine breakthrough after two years of filter operation, while 60% and 40% breakthroughs were observed in the filters that received non-ozonated and ozonated influent, respectively.

- It is difficult to attribute the inaccurate prediction of the PFHSD model to a single cause. However, it can be explained in many ways. For example, the model neglects reductions in the internal mass transfer rate of atrazine caused by desorption of other compounds in GAC filters (*e.g.* brought about by increased water temperature or a change in water quality), while the temperature of Rhine River water during the course of this research ranged from 2°C to 21°C. In addition, the model works with Freundlich coefficients K and n that are constant and, thus, independent of the initial atrazine concentration. This cannot be assumed valid if BOM and atrazine compete for the adsorption on preloaded GAC. It is also possible that the quality of Rhine River water was not the same during the pilot plant operation (1993-1995) and during the preloading of the GAC used to determine model parameters (1996). Finally, seven months of preloading may have been too short a period to reduce the adsorption capacity and mass transfer parameters to the extent required to approximate the breakthrough of atrazine during the two years of pilot plant operation.

GENERAL CONCLUSION

In summary, the following general conclusion can be drawn:

Ozonation significantly improves the removal of atrazine by GAC filtration. This is not only due to the well-known effect of ozone-induced oxidation of atrazine, but also due to the effect of ozone-induced oxidation of a part of Background Organic Matter (BOM) present in water.

Ozone-induced oxidation of BOM improves adsorption of atrazine in GAC filters. Atrazine biodegradation in the GAC filters receiving either ozonated or non-ozonated influent was not demonstrated. However, it was not entirely excluded either. Considering the relative resistance of atrazine to biodegradation, this finding does not imply that other, more biodegradable pesticides are not biodegraded in GAC filters to measurable extent.

The improved adsorption of atrazine is the result of both the higher adsorption capacity of GAC for atrazine, and the faster external and internal mass transfer rates of atrazine in filters receiving ozonated influent. Higher adsorption capacity and faster mass transfer rates can be explained as due to reduced competitive adsorption and reduced preloading of ozonated BOM.

Reduced competitive adsorption and reduced preloading of ozonated BOM can be attributed to the increased biodegradability of an important part of BOM compounds that are partially oxidized by ozonation. Namely, this increases the amount of BOM that is biodegraded rather than adsorbed in GAC filters. Besides increased biodegradability, decreased adsorbability of oxidized BOM was also shown to contribute to the improved adsorption of atrazine.

An accurate prediction of the GAC filter performance is not yet possible. This can be expected considering that, due to the complexity of the processes that simultaneously take place during GAC filtration, the prediction of its performance involves many inevitable assumptions and simplifications.

Samenvatting

Orlandini E. (1999). *Pesticide removal by combined ozonation and granular activated carbon filtration.* Proefschrift, Internationaal Instituut voor Infrastructureel, Hydraulisch en Environmental Engineering (IHE) en Landbouwuniversiteit Wageningen, 171 blz.

Sinds de jaren zeventig zijn nieuwe processen bij de bereiding van drinkwater uit oppervlaktewater geïntroduceerd. De belangrijkste functies van deze processen zijn het zeker stellen van de desinfectie en het effectief verwijderen van bestrijdingsmiddelen en andere organische microverontreinigingen. Gemeentewaterleidingen Amsterdam (GWA) heeft recent twee nieuwe integrale concepten voor de behandeling van Rijn water ontwikkeld. Deze concepten omvatten de voorzuivering van het direct onttrokken rivierwater via coagulatie, sedimentatie en snelfiltratie, gevolgd door een nazuivering via langzame zandfiltratie en hyperfiltratie of via ozonisatie, actieve kool filtratie, langzame zandfiltratie en hyperfiltratie.

In tegenstelling tot de in Nederland gebruikelijke zuiveringssystemen, wordt bij deze twee concepten geen gebruik gemaakt van bodempassage of voorraadvorming. Daarom moeten deze concepten een zodanig robuuste barrière tegen alle mogelijke verontreinigingen in het rivierwater vormen dat onder zowel normale als bijzondere omstandigheden betrouwbaar drinkwater wordt geproduceerd. GWA stelt zich ook tot doel het zoutgehalte te beperken (met name het chloride gehalte) en de levering van onthard en biologisch stabiel water te continueren. In verband met de toekomstige uitbreiding van het zuiveringsstation Leiduin heeft GWA de werking van deze twee concepten getoetst en geëvalueerd. De mechanismen die bij de verschillende processen een rol spelen zijn in het kader van een onderzoeksproject bestudeerd dat binnen een samenwerkingsverband van IHE, NORIT, Kiwa en GWA is verricht. In dit proefschrift beschreven onderzoek vormt een onderdeel van dit project. Het onderzoek is gericht op biologische actieve kool filtratie ofwel de combinatie van ozonisatie en actieve kool filtratie.

De algemene doelstelling van het verrichte onderzoek is vaststelling van de mechanismen die zijn verantwoordelijk voor de verwachte gunstige invloed van ozonisatie op de verwijdering van microverontreinigingen door middel van actieve kool filtratie en zo mogelijk de kwantificering van de bijdrage van deze mechanismen hieraan. Deze kennis opent de mogelijkheid voor verdere optimalisatie van het biologisch actieve kool filtratie process. Een gedetailleerd onderzoek voor de alle in de Rijn aanwezige organische microverontreinigingen was niet mogelijk. Daarom is gekozen voor een modelverbinding, te weten het bestrijdingsmiddel atrazine. Atrazine was gekozen omdat de verwijdering hiervan in 1992 (toen dit onderzoek begon) voor de GWA hoogst relevant was. De concentratie van atrazine in de rivier Rijn was destijds hoger dan de drinkwaterrichtlijn van de Europese Unie (0.1 µg/l). Een tweede reden voor deze keuze was de beschikbaarheid van een analytische methode voor atrazine met de benodigde waarnemingsgrens.

Oriënterende experimenten met een proefinstallatie bij GWA met het via coagulatie, sedimentatie en snelfiltratie behandelde Rijn water hebben de verwachting bevestigd dat ozonisatie de verwijdering van atrazine door middel van actieve kool filtratie sterk verbeterd. Deze verbetering treedt niet alleen op vanwege het bekende effect van de oxydatie van atrazine door ozon, maar ook vanwege de oxydatie van de in het water aanwezige "Background Organic Matter" (BOM). De term BOM staat voor de alle in het water aanwezige organische verbindingen behalve die verbindingen waarvoor actieve kool filtratie bewust wordt toegepast. BOM bevat meestal het natuurlijk voorkomend organisch materiaal maar kan, vooral in de rivier Rijn ook vele verbindingen van huishoudelijke en industriële oorsprong bevatten. De reikwijdte van dit onderzoek is beperkt tot het verwijderen van atrazine in actieve kool filters. Het verwijderen van de tijdens ozonisatie gevormde oxydatieprodukten van atrazine (bijv. desethylatrazine en desisopropylatrazine) is derhalve niet onderzocht. Naar verwachting kunnen deze verbindingen gemakkelijker biologisch worden afgebroken dan atrazine. Anderzijds is tijdens dit onderzoek aangetoond dat deze oxydatieprodukten slechter adsorberen dan atrazine. Daarom is het moeilijk te voorspellen of de oxydatieprodukten van atrazine in actieve kool filters beter of slechter dan atrazine worden verwijderd.

Het doel van de uitgevoerde experimenten met proefinstallaties en laboratorium opstellingen was vast te stellen welke van de verwachte mechanismen en verbanden ten grondslag liggen aan de waargenomen verbeterde verwijdering van atrazine in actieve kool filters met geozoniseerd influent. Tijdens ozonisatie wordt namelijk een deel van BOM gedeeltelijk geoxydeerd. Dit verhoogt de biologische afbreekbaarheid van BOM verbindingen, terwijl het hun adsorbeerbaarheid op actieve kool en hun moleculair massa verlaagt. Daardoor wordt verwacht dat in filters gevoed met geozoniseerd water de biodegradatie van het BOM wordt verhoogd en de adsorptie wordt verminderd. Beide effecten kunnen resulteren in een versterkte biodegradatie en een verbeterde adsorptie van bestrijdingsmiddelen.

Op basis van de uitgevoerde experimenten wordt verwacht dat de verdergaande verwijdering van BOM in filters gevoed met geozoniseerd water, ten opzichte van met niet-geozoniseerd water gevoede filters, wordt veroorzaakt door de verhoogde biodegradatie van BOM. Deze conclusie is gebaseerd op het feit dat de verwijdering van geozoniseerde BOM in filters gevuld met niet-actieve kool (waarin BOM via biodegradatie verwijderd wordt) beter verloopt dan de verwijdering van niet-geozoniseerde BOM in deze filters. Er kon echter niet worden aangetoond dat biodegradatie van atrazine een rol speelt bij de waargenomen verbeterde verwijdering van atrazine in actieve kool filters die gevoed zijn met geozoniseerd water. Een indicatie voor de biodegradatie van atrazine is namelijk niet gevonden in een van de experimenten die hierop gericht waren, te weten: (i) er zijn geen metabolieten van atrazine gevonden in het effluent van actieve kool filters waarin atrazine was toegevoegd, (ii) atrazine werd niet verwijderd in de filters gevuld met niet-actieve kool, (iii) atrazine werd niet verwijderd in de vloeibare media geënt met de bacteriën uit de actieve kool filters die gevoed zijn met atrazine houdend water, en (iv) er kon na drie jaar bedrijf meer atrazine worden gedesorbeerd uit de filters die zijn gevoed met geozoniseerd water dan uit de filters die zijn gevoed met niet-geozoniseerd water. Verhoogde biodegradatie van BOM in filters gevoed met geozoniseerd water verbetert adsorptie van atrazine in actieve kool filters. Dit wordt geconcludeerd omdat atrazine beter was geadsorbeerd op actieve kool die was voorbeladen met geozoniseerd water dat over niet-actieve kool filters was geleid (in deze filters wordt BOM verwijdert via biodegradatie) dan op actieve kool die rechtstreeks met geozoniseerd water was voorbeladen.

De verkregen resultaten hebben ook aangetoond dat verbeterde adsorptie van atrazine in de filters gevoed met geozoniseerd water ten opzichte van de filters gevoed met niet-geozoniseerd water door zowel de hogere adsorptiecapaciteit voor atrazine als door de snellere externe en interne stofoverdracht van atrazine kan worden verklaard. Een hogere adsorptiecapaciteit en een snellere stofoverdracht van atrazine kunnen door de gereduceerde concurrerende adsorptie en de gereduceerde voorbelading van de geozoniseerde BOM worden verklaard. Concurrerende adsorptie van BOM vindt plaats als BOM en atrazine tegelijkertijd adsorberen en daarom voor de op actieve kool beschikbare adsorptieplaatsen concurreren. Voorbelading van BOM betekent dat BOM wordt geadsorbeerd op actieve kool voordat atrazine daarop adsorbeert. Gereduceerde concurrerende adsorptie en voorbelading van BOM zijn onder andere het gevolg van de hogere biologische afbreekbaarheid van een deel van BOM verbindingen die door ozonisatie gedeeltelijk zijn geoxydeerd. De verbeterde biologische afbreekbaarheid verhoogt de hoeveelheid BOM die in actieve kool filters biologisch afgebroken, in plaats van geadsorbeerd wordt. Niet alleen de verhoogde biologische afbreekbaarheid van geoxydeerde BOM verbindingen maar ook hun verminderde adsorbeerbaarheid bijdraagt aan de verbeterde adsorptie van atrazine in de filters gevoed met geozoniseerd water ten opzichte van de filters gevoed met niet-geozoniseerd water.

Tot slot zijn twee algemeen toegepaste modellen te weten het eenvoudige model van Adams-Bohart en het Plug Flow Homogenous Surface Diffusion model toegepast om doorbraak van atrazine in actieve kool filters te voorspellen. Dit zowel in filters werkend met en zonder ozon gestimuleerde bioactiviteit. Geen van de twee modellen heeft in een nauwkeurige voorspelling geresulteerd. Een dergelijke bevinding is echter te verwachten vanwege de complexiteit van de processen die tijdens actieve kool filtratie gelijktijdig plaats vinden. Daarom vergt het gebruik van de modellen onvermijdelijke veronderstellingen en vereenvoudigingen.

Sleutel woorden actieve kool, adsorptie, atrazine, Background Organic Matter, bestrijdingsmiddelen, biodegradatie, biologische actieve kool filtratie, bromaat, concurrerende adsorptie, disinfectie, modellering, ozonisatie, procesanalyse, voorbelading.

Curriculum vitae

Ervin Orlandini was born in Split (Croatia) on February 3rd, 1962.

He received B.Sc. degree in Civil Engineering from the University of Split in 1986 and M.Sc. degree (cum laude) in Water Quality Management from the International Institute for Infrastructural, Hydraulic and Environmental Engineering (IHE) in 1992.

From 1986 to 1991 he worked for the Water Authorities of Croatia as a project engineer dealing with the collection and treatment of municipal wastewater. He was certified as Professional Engineer in Croatia in 1988. From 1992 to 1997 Mr. Orlandini was a researcher at the IHE, working within the field of drinking water technologies. Since 1998 he is with Grontmij Consulting Engineers as a consultant for drinking and process water.